BIKE SQUAD RIDER

Yuri Maree

ISBN 979-8-218-19010-1

bikesquadrider@gmail.com

In Memory of Patrick Devy
1 June 1961 -- 5 May 2019

Introduction

The bush war in the former German colony of South West Africa, Namibia today, started in 1966 as a consequence of the 'winds of change' that blew through Africa in the 1960s. Also known as the Border War, the small-scale hostilities were officially classified as a 'police action' until the SADF's involvement in the Angolan civil war in mid-1975. A South African battle group, largely armed with Second World War weapons and equipment, joined the FNLA and UNITA movements in their war against the Marxist MPLA which had seized power after the Portuguese colonial government withdrew and left the country to its fate. Clandestinely supported by the USA and its CIA, a unit known as Bravo Group advanced to the hills overlooking the capital Luanda by November 1975 but withdrew after the US government halted aid to UNITA and FNLA in December. Large numbers of Cuban personnel under Soviet command were imported to prop up the Marxist MPLA government.

UNITA retreated to southern Angola along the border with South West Africa, which was governed by South Africa under a 1919 League of Nations (later UN) mandate. The South West Africa Peoples' Organisation (SWAPO), also supported and armed by the Soviet Union, saw an opportunity to 'kick the colonials out' and a counter-insurgency war ignited that lasted another 13 years. Having learned some harsh lessons in 1975, the SADF was extensively re-equipped and reorganised. Clandestine involvement in the Rhodesian bush war taught the SADF valuable lessons about counter insurgency operations and it effectively destroyed the Peoples' Liberation Army of Namibia (PLAN), the military wing of SWAPO, by 1985.

The MPLA's Cuban army and air force threatened to invade South West Africa and by 1988 the war escalated into tank battles and aerial combat on a scale last seen on the African continent during the Second World War. South Africa started full mobilisation, and the dying Soviet Union could no longer support its proxy forces in Angola. Cooler heads prevailed on both sides and the Border War ended in 1989. South West Africa became independent as the country of Namibia in March 1990.

During the early COIN phase of the war between 1976 and 1979 the SADF experimented with many innovative operational practices. Analysis of previous COIN wars (like the British experience in Malaya) and experience gained from its secret involvement in the Rhodesian bush war led to the correct conclusion that insurgencies can only be defeated by 'boots on the ground'. New doctrines were developed, and physical tracking of insurgents became a priority.

The original inhabitants of southern Africa, the San people known as the 'Bushmen', had been overrun by the expanding war in Angola and were left to the mercies of their historic enemies -- black Africans – by the Portuguese government when it fled the country in 1975. When Bravo Group withdrew from Luanda in December 1975 and early 1976, it evacuated two tribes of Bushmen at risk of extermination to South West Africa. They were eventually employed as soldiers by the SADF, which recognised the value of their tracking skills and bushcraft.

A specialised tracking programme was developed to copy what was learned from the tough little men who could tell the age of tracks or 'spoor' (the Afrikaans word) to within an hour or two, whether the quarry was tired or fresh, and how heavy their packs were. Tracker dogs were trained and formed into a separate unit, and horse- and motorcycle mounted specialist troops into yet another. All these specialised tracking functions except the Bushmen were combined into 1 South West Africa Specialist Unit (1 SWASPES) on 1 June 1977.

The relatively open, absolutely flat terrain in northern SWA, combined with long distances, soon proved ideal for motorcycles. The highly mobile platoons were extremely effective during quick-response follow-up operations, and PLAN's insurgency effectively ended in 1982, the year chronicled in this memoir. From 1983 on, tracking became less of a priority and the motorcycle platoons were used in any role their extreme mobility was useful. By that time unrest was breaking out in South Africa itself and dogs, horses and motorcycles were trained and deployed in urban environments, specifically Soweto near Johannesburg, from 1984 on.

Bike Squad in the form described in this book only existed for eight years. Less than 500 men rode into action like motorised cavalry, a unique function which no other force did before or since. German units on the Eastern Front during the Second World War came closest to the model employed by SWASPES from 1977 to 1983, but did not ride into contact as Bike Squad did. All other forces, including the US in its wars in Afghanistan and Iraq, used motorcycles as transport, not combat vehicles.

SADF motorcycle doctrine was surprisingly well thought out and was only modified slightly under operational conditions, none of it official. The most obvious deviation from doctrine was the lack of HQ sections in operational platoons, as indeed in all SADF counter-insurgency forces. The full HQ section containing medics, signallers, sappers and other support services was conventional warfare doctrine that did not work well under the more dynamic conditions encountered during COIN ops.

The biggest deficiency in SADF motorcycle doctrine was the lack of continuity and development. The operational platoons developed many habits and practices which were never incorporated into official doctrine, or even transferred to successive platoons. Attaching specialist units like motorcycle and horse platoons to ordinary infantry units whose commanders were trained in conventional and COIN infantry doctrine, wasted much of the unique capabilities of platoons that could outrun everything except helicopters in Ovamboland and southern Angola. As described in this book, company and battalion commanders who knew nothing about the strengths and weaknesses of motorcycles and had an ingrained dislike of 'bikers', never understood how much of a force multiplier the extremely mobile and motivated motorcycle platoons were.

During its eight year existence, Bike Squad was never more than two platoons strong. Some overlap with July intakes of one platoon did occur, but made no impact on operations. After 1985 the Potchefstroom home-based motorcycle unit was diluted and used in ways motorcycles had been since the First World War, often as dispatch riders and light transport. A 'local' SWATF unit named 2 SWASPES was founded in the Operational Area, but very little is known about it except that it had at least one fatal crash on the tar road that diagonally traversed Ovamboland.

1983 was the high point for the 'mechanised cavalry'. An intensive three week testing programme was conducted at the GEROTEK testing ground west of Pretoria and the Riemvasmaak training area near Upington, to select the SADF's future operational motorcycle from eight different types submitted by five different manufacturers. The author of this book was involved in a command capacity but had no decision-making powers. After much effort and money spent, the Honda XR 500R remained the motorcycle of choice but it was a moot point. After PLAN was destroyed by 1985, the bush war became more and more conventional until the escalation made both sides blink and the war ended in 1989.

At the time of publication in 2022, the SANDF employs motorcycles ostensibly as the SADF did 40 years ago, but the reality is far from the unique character of the original Bike Squad. It is safe to say, the likes of Platoons 1 to 14 will never be seen on the African plains again, or anywhere else.

TABLE OF CONTENTS

TABLE OF CONTENTS

1
There I Was...

...getting on with life, and wondering why the tall palm tree in my neighbour's back yard made my pulse race every time I saw it silhouetted at dawn. It made me laugh too, which was baffling as trees are not generally amusing. It took me a few years to connect the dots.

As time went by, memories of my year in the bush war with Bike Squad hovered around the edges of conscious thought. I'd gone walkabout in 1984 after leaving the SADF, and the challenges of living in foreign countries and pursuing a difficult career path allowed only occasional brief reminders. I never stood around braai fires telling *grensvegter* stories. Then, in 2006, an unexpected whiff of decomposing meat in a dustbin erased a quarter of a century and suddenly it was 1982 again.

Old memories became a big part of my day for a while and I dug out my photos from that year. I'd looked at them once in a blue moon, usually when I stumbled upon the box while looking for something else. A handful were seen by a few friends when conversations turned to motorcycles, but they mostly lay forgotten for more than 20 years. To preserve the photos before they faded, I scanned them onto computer disc. I was able to enlarge them enough to make many long-forgotten details jump out at me.

I decided to record the stories behind the photos while I relived them. Once I started, my fingers couldn't keep up with my brain. The overwhelming detail that came rushing out made me wonder if I was twisting the truth or making things up. Two independent sources reassured me that I wasn't. I'm still close friends with several of my old bike squad connections, one of whom was with me at SWASPES for the first part of that year. They all confirmed details and provided bits of information that told me I wasn't getting creative with the facts.

The second stamp of validation came from the website Google Earth. It's an interactive display of satellite photos in global form, detailed enough that I can see my white bakkie standing in my driveway. While playing with this amazing toy one evening, I tried to find some of the locations where specific incidents happened.

I found many. They were exactly as I remembered them.

The stories that follow are my recollections of events and incidents I was involved in or observed. They're not all I remember, just some of the more amusing and memorable. My old platoon mates may recall them in less or more detail, or not at all. Except for a few swapped with old army mates, I have not told these stories before.

Details like dates, places and distances escape me to a large extent. I made notes on the backs of many photos at the time, without which I'd have no clue of the context. The names mentioned in the stories are real. Quite a few of them are dead. The names of most of the individuals we despised, I never knew or don't remember.

Because of forgotten chronology this is a collection of short stories. I remember 1982 that way, much of it as scenes captured by my little 110 camera. The stories are written in the parlance of the time. The SADF was profane and so were we. I cleaned up the language considerably but it's NOT family-friendly, censored, or politically correct. If you don't like it, don't read it. Snow White and the three fucking bears this isn't.

I spent 1982 'on the border' as platoon sergeant of bike squad platoons 11 and 12. I'm sure it wasn't always obvious to those okes – I was the hated and feared corporal, after all – but I was very proud of them and remember them fondly and with great respect.

I hated much of the four years I spent in the army for the same reasons Roman legionaries in Carthage and muddy Tommies in Somme trenches did, but 1982 was one of the best years of my life. Despite the boredom, the danger and the kak food, I really miss those days lately. The days of being an adrenaline-fuelled 20-year-old riding dirtbikes all day not knowing, or caring really, if it would be the last thing you did in this life.

Part of me is still wheelying across shonas and jaaging through mopani bush. At times, I wish all of me were.

On our first patrol in Angola, note anthill leaning north.

*South of Oshakati. In this kind of terrain bikes outran
everything except helicopters.*

2
Windows

Looking back on life, one can clearly identify pivotal moments through the objective lens of time gone by.

One such moment actually lasted a few seconds but had a huge impact on my life. Paying attention to a question asked of newly hatched infantry corporals in a queue ahead of me brought the most enduring constant into my life. I had joined the PF with the intent of perambulating about the skies in a jet, but that scheme was shot down. A long story, the moral of which was don't get pissy when an idiot asks you the same question three times. Especially when the idiot is a Colonel, and you're a *troep*.

Consequently, after a year of *afkak* and *rondfok* at Infantry School, I realised that some serious innovation was needed to avoid marching around like a *bokkop fokkop* for three more years. I applied for, and got, an assignment to the SADF Equestrian Centre in December 1980. Heroic images of horse-mounted warriors had entered my brain somehow, and I could see myself leading a modern version of the Charge of the Light Brigade. I would survive, of course, to bask in glory and the charms of beautiful young *pundas* far and wide.

I don't know what the fuck I was thinking.

I'd been on a horse twice in my life and fell off both times. My newly acquired dislike for running around with a steel pot on my head and long hikes with heavy packs probably eroded what little good judgement I had. Once away from Oudtshoorn, I realised that. The closer I got to joining Berede the more nervous I got. By the time I reported for duty after a three week holiday I wanted fuckall to do with anything on four legs.

A dozen new Corporal instructors arrived at Berede in January 1981. Four were going to bikes and the rest to horses. The bike squad contingent were all from my company at Oudtshoorn. Two of them had been in my platoon, and we quickly renewed the bonds forged in hardship over the previous year. After a formal welcome ceremony in the thatch-roofed PF tea room that first morning, the laughing, joking group of Corporals queued in front of a clipboard-wielding Sergeant for assignments. I was prepared to grovel, beg or kiss arse to go anywhere but the stinking horses. Knowing that I was there for the army's benefit and not mine, I did not view my immediate future with enthusiasm.

But, as I got closer to that clipboard and a dreaded three-year sentence, I realised something about the army.

Again.

The left hand often didn't know what the right hand was doing. *'Twixt the cup and the lip, the pap falls in the dust.'*

The stocky little Sergeant with the fu-man-chu moustache and clipboard asked the Corporal in front of me: *'...what are you? Berede or bikes?'* It only took me a second to catch on. The unit's personnel officer had been sent a name list of new Corporals, with no mention of the assignments made at Infantry School. Some lazy clerk back in Oudtshoorn changed my life.

I marched up to the blonde Sergeant with the sloping forehead and came to a thunderclap halt in front of him. I had *houding*, you see. In his most *paraat* PF voice, modulated to convey discipline, leadership and organisational skill to the RSM's ears lurking nearby, he asked me the big question.

'...Bikes Sergeant' was all it took.

Twenty-six years later I still laugh about it every time I ride one of my fifteen motorcycles.

3
Ballas the Beagle

One of the chaplains at SWASPES was a two-pip loot who lived with his wife and little kid in the school that served as officer and NCO quarters. He had a dog that we named *Ballas* for two very obvious reasons the first time we saw him. Ballas was a Beagle. My sister had one a few years ago, so in hindsight I understand Ballas' behaviour a little better.

Ballas was a source of much amusement for me and my berede platoon sergeant friend Ikes. He was friendly, and seemed a bit *dof*. He had a black dot between and slightly above his eyes, which we couldn't resist pressing like a doorbell. It made Ballas wag his tail and smile like dogs do when they get human attention. We soon progressed to pressing on his dot and when he looked at it cross-eyed and wagged his tail, we'd cock back our middle finger and thumb and pop him on the nose. It made him jump big time. After we did this to him four or five times he started growling at us. It got to the point where he'd almost take your fucking hand off when you touched his dot, so we labelled it 'panic button'.

It became our new game, trying to sneak up on Ballas and poke his panic button, then jump away and laugh our arses off as he went berserk. Eventually he started growling and snapping his teeth as soon as he saw Ikes or me, and he'd walk backwards to get away from us.

Beagles are described as '*single-minded and determined*' by dog experts. I do suspect Ballas was different from the rest of his kind after making our acquaintance. Poor bastard dog probably had emotional scars the rest of his life.

4
My Favourite Number

The RSM at SWASPES was a funny oke. He was dark-haired and balding, wore Buddy Holly glasses and sported a *boep* years in the making. He was very *rustig* for a RSM. He yelled at times to keep up appearances, but really didn't seem as nasty as some other bastards of the breed. He was also no Einstein. Rumour had it he only finished standard six, but overall, he was a decent human being.

Word got out among the NCOs that if you fucked up in some way, the RSM would engage you in random conversation and slip in a question about your favourite number. Whatever you said, he'd then give you that many 'extras'. For us lowly-but-godly Corporals, 'extra' meant being the night Duty NCO which we all hated. It was a night of interrupted sleep and we considered it rondfok. The pukka gen was, learn to like single-digit numbers.

Came my turn to be night Duty NCO at Spes. At dawn I stood on a covered walkway at the school that served as HQ, waiting for 08:00 so I could hand over to the day shift and go chase swaps. I was content. Despite gyppoing most of the night, nothing had happened that I'd have to go explain to somebody. Then the RSM wandered up and stood next to me. He started making small talk and I instantly became suspicious.

Sure enough, he proceeded to point out to me that I had not lowered the flag, ten metres away, the previous evening. I'd forgotten and went to bed early but I didn't tell him that. I splabbed some insincere apology and he casually asked me what my favourite number was. Battling to keep a straight face, I said '*..zero, samajoor*'. It took the wind out of his sails big time. He *hum'ed* and *haa'ed*, coughed, shuffled his feet, and said something like '*...uuumm ...you'll have to do some extras*' and walked away. I couldn't tell whether he was amused or confused, but I started laughing as soon as he was out of sight.

Whichever it was, I never did another Duty NCO shift at SWASPES.

View of SWASPES, looking north. Bike squad and berede offices left foreground, the mess hall behind.

Alouette gunships at SWASPES, April 1982.

5

To Sleep Perchance To Dream

During Ops Yahoo, platoon 11 took part in a big stopper group across the suspected route of some *Volcano* terrs one day. It was a 61 Mech operation with about 20 Ratels. The turbo-diesel engines in the Ratels made a very distinctive and loud whining sound. I wondered how the fuck we were supposed to sneak into an ambush position with all that noise going on.

The Ratels left Spes painfully early at 03:00. We were cold, tired, and packed in like sardines. Many okes went back to sleep. The section I was with brought their pet baby baboon along, so we entertained ourselves playing with it. Rifleman Kruger, the platoon's scrounger and a dedicated shit-stirrer, as usual didn't know when enough's enough and teased the baboon till it attacked him. Stupid move. Baboons may appear friendly and cuddly like your homely spinster aunt, but they are extremely vicious when provoked and those long arms can literally rip your head off.

Kruger got off lightly. This one was maybe 20 cm and two kgs and clung to its owner while staring daggers at the tormentor. Only a few weeks old, he had all the devious and vicious instincts of his kind. I was watching his face and saw it coming: he got a fierce, wild look in his eyes and launched off the oke he was hanging on to. In one flying leap he was on Kruger, screeching and biting and scratching at him. Kruger screamed at the others to get the baboon off of him, but we were laughing so hard we couldn't do anything. Besides, we schemed he deserved it.

The ambush was placed just inside the bush, on the south side of one of those distinctive snow-white roads about 70 kays south of Otavi. The Ratels were spaced over a distance of two kays with troops in between. The dense bush had been bulldozed clear 10 metres on each side of the road, which created a kill zone 30 metres wide. The blade berm left along the treeline by the bulldozer provided perfect cover for the troops lying in ambush.

Sunrise came and went. No SWAPO. Soon after, the *trek af plak toe* command was given and we climbed into the Ratels and fucked off back to Spes. Minutes after arriving a troop came and told me that Corporal Russell, one of the section leaders, was missing. We immediately sent two okes on bikes back to the ambush site to look for him. Some two hours later they returned with a grinning Russell, very happy to see his mates again.

He had fallen asleep and slept right through the noise of 20 Ratels starting up and leaving. He had quite a shock when he woke up alone in the bush, not a soul in sight, and some very bad terrs coming at him as far as he knew.

An easy going blonde oke who smiled a lot, Russell was teased about his sleeping habits for a while but he just shrugged and lagged it af. The less amusing part was, a Ratel had missed him by half a metre when it drove off. They found the tracks right next to where he'd been sleeping.

6
Young Man's Game

My platoon in the bush was typical of most infantry units, I suppose. After eighteen months in the army, nine on ops, the youngest troop was sixteen and the oldest twenty-two. I remember three okes who were twenty-one, the rest were all nineteen. The loot turned twenty in Angola not long before the platoon klaared out. I turned twenty that year in the back of a Samil 50 on the way to a military funeral in Keetmanshoop.

Part of platoon 12 in Ovamboland displaying the haircuts the 2IC at Ogongo forced on us. Back row, L to R: Banks, Sutton, Heath, Douglas, Vorster, Steinvaart, Geeringh. Middle row, L to R: Ingalls, Beudeker, Le Roux, HAPS du Plessis. Front row, L to R; Welsh, Reeves, Van Rensburg, Maree.

7

Hunters Are Wankers

During SWAPO's farm incursion, I led two platoon 11 sections on a crack-of-dawn follow-up in two Buffels one cold morning. We did lots of things before sunrise during that time. We left SWASPES in the dark, I was in the lead Buffel with a local farmer as guide. The farm where the terrs had caused *kak* during the night was about 80 kays south-east of Otavi, near the mountain range that's a nature reserve today.

The guide, *oom* Dirk, had spent several days with us and told us about three cheetahs the farmers in the area had been hunting for several weeks. They were on the shitlist for killing cattle and sheep. While chasing swaps we checked out a few waterholes and a cave where the *oom* thought they might be, but had no luck.

We came flying around a bend not far from our destination and suddenly three cheetahs were running in the road right in front of the Buffel. The farmer yelled '*…skiet hulle, skiet hulle!*' as the cheetahs stopped and turned and ran back the way they came, right past us. The driver stopped, and one of the troops standing behind me on the left side was first to react.

He pointed his R4 straight down over the side and shot the leading cheetah. In the confusion of the moment he had it on automatic, and must have emptied a whole magazine into that poor cat. Red tracers flew all over the fucking place. It was just after sunrise and most of the troops were half asleep. They were rudely awakened and popped up above the Buffels' high sides with big eyes and R4s in hand.

Meanwhile, the other two cheetahs had run maybe 70 metres behind us and angled away from the road on a farm tweespoor into the bush. I stood up on the seat, aimed diagonally across the Buffel over the troops' heads, and shot the rear one on the run. It tumbled in a cloud of dust. Our farmer guide was extremely impressed. He thought I was hero of the day and babbled on and on about my marksmanship.

We went over to the cheetah, a female. My shot had gone through her shoulders and spine. Despite being paralysed, she was still snarling and lunging at us. Quite a scary sight from a metre away. I shot it in the head twice, we threw both dead cheetahs in the back bin on the Buffel and drove to the farm house nearby.

The two farmers were very happy about the diminished threat to their cattle. Our guide was still rattling on about my shot that got the second one, but I was feeling less than heroic about it. The farmer said he would skin and preserve them. We saw the skins later, on another visit to the same farm. One looked like Swiss cheese but mine was in good nick. It only had a hole on each side through the shoulders and two in the head. The guide nagged me to go back and get 'my' skin but I just couldn't make myself do it.

We found lots of SWAPO *alpha-sierra* spoor at the workers' houses and I took photos of it. We tracked it for a while but eventually lost it in the harsh, rocky terrain. I didn't find out till several days later that I almost got in big shit for shooting a protected animal. Our shots had carried a long distance in the cold, still morning air and several farmers called the HQ in Otavi in a flap and reported a contact. Somehow the nature preservation people found out about it and caused a stink. The CO of the Otavi Commando who was in charge of the operation and also a local farmer, squashed it.

I've always felt sorry about shooting that cheetah. It was a beautiful animal whose only crime was having its habitat invaded by humans. I despise people who shoot animals for the skin or horns. Hunting for food is fine, for a head on a wall is not. I'd like to put trophy hunters out in the bush and track them down with a rifle in my hands, like we did SWAPO.

Fucking wankers.

8
Happy Birthday

I spent my twentieth birthday in the back of a SAMIL 50 droning between Otavi and Keetmanshoop. Those trucks, magnificent pieces of licensed German engineering, were governed to 80 km/h which made it a day-long ordeal with a stop in Windhoek. Herbert the German loot, my friend JB and I had been detailed to do a military funeral with a platoon 11 section. The dead soldier was being buried in his hometown of Keetmanshoop, the biggest town in the southern part of the country.

An image of dry yellow grass along the arrow-straight tar road, framed in steel and canvas, was burned into my memory because it didn't change for hours on end. I also remember sitting in a Keetmanshoop pub with hideous aquamarine walls. The evening sun was shining in through a tall window with a lace curtain blowing in the wind, and reflecting off the dark wood bar counter. My two friends were sitting up-sun on my left. JB was pontificating about Scotch whisky being an acquired taste, so I tried a single on ice just to shut him up. To my great surprise, I liked it.

The funeral was the day after my birthday. I can't remember a thing about it. Not the slow marching I know we did because we'd practiced for hours, not the grave site, not the people or the coffin. Not where we stayed, or the trip back to Spes. Not a fucking thing except that the oke we put in the ground was a year younger than me.

9
Pro Patria Mori

Every year around March when the wet season came to an end, SWAPO sent highly trained terrs with bad intent to the farming areas around Tsumeb and Otavi. My bike platoon was part of the months-long operation to track and wipe this lot out. SWASPES trackers, 61 Mech Ratels, parabats, Koevoet, and Alouette gunships made up the rest of the force. Those swaps were very tenacious, we had no illusions about them. Because of the bikes' mobility, the platoon was used as a quick reaction force. We mostly did follow-up operations at high speed on short notice.

One hot and dusty afternoon the loot and I were sitting in the shade across from the Commando HQ in Otavi, waiting for something to happen. One of the parabat officers, Captain van Wyk, came walking out of the HQ across the street towards us. We both stood up from where we were comfortably lying back in our webbings. The Captain waved *'don't bother'* and stopped to talk. We mixed with the parabats in the mess and pub and were on friendly terms, so the three of us chatted for a while. Captain van Wyk said cheers and walked away. We sat back down to ballas bak a bit longer.

Four hours later, a terr cornered in a hut a few kays away by the bats and 61 Mech threw a grenade that killed the Captain almost instantly. We heard the news in the mess at brunch the next day. I don't recall having much of a reaction, but twenty-five years later I still think about him. I can still see him standing in that dusty street. His face is a little blurry, but he is smiling.

10
Comrades

Not much propaganda was put out during the bush war, it was a very secretive affair. The only thing from our side that I recall, was hearing *skyshout* a few times. It was a light aeroplane flying up and down the cutline at night with powerful speakers talking *kak* in Ovambo. Or maybe Portuguese, I couldn't tell.

We picked up some SWAPO propaganda booklets printed on blotting paper, like we finger-painted on in primary school art class. They were full of black-and-white photos with red text and clenched fists, honouring the *comrades* who had fallen in the *struggle*. The *comrades* were dead swaps in all kinds of gruesome poses. Half a head, no arms, blood and guts type of shots, with text insulting the *boere* and telling heroic tales of battles won.

I found it very amusing. Maybe I'm dense, but I never quite understood how photos of your own mutilated dead were supposed to motivate the living to fight.

The pamphlets were printed on paper too rough to wipe your arse with, but we soon found a linguistic use for them. At first, *comrades* meant 'dead swaps'. One day someone blew the arse trumpet and the loot disgustedly said that it smelled like comrades. That soon evolved into '...*what the fuck did you eat? comrades?*' when somebody stank the place up. And that is how *'eating comrades'* came to mean *'shitting your lungs out'*.

The live comrades we encountered during Ops Yahoo were anything but amusing. They did not achieve their objective of terrorising the local civilians but they kept us, and many others, very busy chasing them. They also killed at least two soldiers in our operations area that I knew of. At the time we were not too familiar with the big picture, we just knew that we were up against SWAPO's elite terrs. That was proven the one day we caught up with three of them.

My platoon, a Koevoet group, and two Alouette gunships chased them from sunup to sundown. I don't remember much about the morning. We acted as a mobile stopper group for the Koevoet team that tracked them through the dense bush around Otavi. The chase started early, the terrs ran west and then turned south. Late that afternoon they turned east again. We leapfrogged ahead to cut them off numerous times during the day but they eluded us every time. Koevoet was literally a hundred metres behind them at times.

I remember lying in wait, watching the gunships circle maybe half a click away. One of the pilots, we were told, was Captain Arthur Walker who had quite a reputation and two Honoris Cruxes. We listened in on the radio exchanges between the gunship and Koevoet. The pilot yelled '...*you guys better fucking move down there!*' as we watched him circle low over the bush. That was impressive. Saying *fuck* on the radio scored some serious points with us.

During the day we were told that Koevoet was finding syringes and needles on the spoor, proof that the swaps were using drugs. I had heard such stories before and took it with a lot of salt. It wasn't till many years later, when I was much better informed about life in general, that I realised they must have injected themselves with steroids or similar shit. That detail was conveniently left out at the time, but either way there were no anti-doping rules in the game we were playing. By mid-afternoon the three had run over 80 kays since sunrise.

Just before sunset we were waiting on a dirt road parallel to the main road about 80 kays south of Otavi. The gunship was circling close by and the pilot radioed that the swaps had turned east. Looking at the map, we saw that the road went south for a bit, then turned east towards the tar road between Otavi and Otjiwarongo. We jaaged ahead one more time and found a fenced cutline with a tweespoor, running north-south across the terrs' direction of movement.

We rode north on this until we came to a big power line that crossed the track at a sharp angle. The power line was also fenced in, with gates on both sides where the tweespoor crossed it. We positioned ourselves in a line along the fence and faced west. I had everybody lie their bikes down so they wouldn't stick up above the tall grass outside the fence. The sun had just set, the pale blue sky contrasted with the dry yellow grass and dark bush that grew right up to the wire fences. Things were a bit tense.

My position was at the end of the line, by the gate at the intersection of the two cutlines. I could see down the powerline to my right and the tweespoor to my left. We hadn't been there five minutes when the troop right against the gate said: '...*hey corporal, what's that?*' and pointed down the powerline. Maybe 200 metres away, what looked like three cows were walking across the open space. Two were in the lead nose to tail, then a gap and one more. It took me less than a second to realise it was the three terrs crawling across the cutline on their hands and knees. I yelled:'...*it's them, shoot!*' and did so, aiming low. The troops close by joined in. The terrs seemed to make it across despite our best efforts.

The gate in the fence had a big lock on it and I started cutting the five steel wires next to the gate with the wire cutter on my R4. One of the troops helped cut the wires and the others mounted up. As soon as we had a gap in the fence, I jaaged down the powerline with the platoon strung out behind.

We passed the first power pylon and I realised that I had fucked up. While shooting at the bastards, my mental picture had been of them being two or three pylons from our position. I now realised I hadn't counted how many, so I had to judge the distance by recalling the sight picture I had while shooting.

When the visual looked right I stopped, we dropped the bikes and spread out in a line facing the dark bush where the three terrs were hiding. The loot must have called contact on the radio because a gunship was hovering overhead almost immediately. I looked straight at the pilot no more than 20 metres away and frantically pointed into the bush. He flew in that direction and circled a few times, but couldn't see anything as it was almost completely dark by then.

After some discussion we decided that a sweep into the bush would result in getting someone killed for no return on the investment. The three terrs had all the advantages. They were lying in the dense bush waiting for us, and the light was fading by the second. They were dead men walking and had fuckall to lose at that point.

There wasn't much else to do. In pitch dark we backtracked down the powerline to the dirt road, headed east for a few kays and got back on the tar road. The temperature had started to drop as soon as the sun set, and it was very fucking cold by the time we regrouped at the road. We were 80 kays from SWASPES and had two options: cruise at slow speed to keep the wind chill down, which would take at least an hour, or *steek* it and freeze for 40 minutes or so.

We hadn't eaten since before sunrise and we were tired, hungry, and the moer in about nightfall denying us three kills. So we did what any self-respecting biker would: we tucked in, put our chins on the tanks of the bigwheel XR 500s, and wound the throttles as far as the cables would stretch. The hopelessly inadequate headlights illuminated only about 20 metres ahead of each bike. At 120, that wasn't much.

About twenty minutes from home we almost lost half the platoon. The loot was in the lead and had pulled some distance ahead. I don't know if he decided to wait for us, or was freezing and needed to warm up a bit. Either way, the image burned into my brain is of him frozen in my bike's puny light beam, standing on the left side of his bike just left of the stripe in the tar road as I missed him by centimetres at 120 plus. When I turned around and rode back to him he was still standing there, white as a sheet. The platoon had flashed by him on both sides with throttles twisted to the stop. I'd had a bit of a *skrik* and yelled something like '...*what the fuck are you doing?*" but it was a moot point and we continued on after we stopped shaking.

We went back to the contact site the next morning. The wire fences were down where Koevoet had driven through them at first light, chasing the spoor. We had been in the right spot after all. Those terrs had run over 90 kays in one day. I remember thinking, they ran a Comrades Marathon in the bush. The irony didn't strike me till many years later.

Once in a while I still contemplate the '*if onlys*' of that evening. *If only* the winter sun hadn't gone down so early. *If only* we had laid that last ambush five minutes earlier. *If only* that bloody gate hadn't been locked. I know it's futile, but I still get a bit miffed when I think about it.

Koevoet killed those three the next morning. They should have been our kills.

11
Vleisbomme

Kevin, my loot friend at SWASPES, told us a story in the Otavi Hotel pub one evening. Between the last time I saw him at Berede in early 1981 and again at Spes a year later, he went through a few career-enhancing courses in the SADF. Like *'How to Bone Yer Boots Extra Shiny'* and parachute training. He never wore jump wings so it may have been a staff course, I forget the details.

Parabats were something special, if a bit *dof*. They were infantry twats, like us, but they thought jumping out of perfectly good aeroplanes was more fun than riding bikes. Selected from volunteers at the SAIs, like us, they went through a Darwinian two-week selection process called *PT Course*. The bats were okes who liked PT. I had turned *nafi* after a skirmish with some wanker Colonel deselected me out of the SAAF, so bats and recces were not part of my career plan. Lost my desire to suffer, pure and simple. Running 21 kays with a fucking telephone pole was not my idea of a good jol.

In the pub that night, Kevin told us a story about being Range Officer one day. Standing in the drop zone with a radio, coordinating a training drop outside Bloemfontein. Everything went fine until one oke had a roman candle. Kevin said the baby bat jumped from 800 feet, the standard drop height, and his chute streamed behind him as he fell. Some angel was watching over his sorry arse because he bounced, shook his head a few times, and walked away.

Kevin swore it was true and miracles do happen, so I believed him. He could have used luck that good on a lonely road through the bush a few years later. No question about it, *jou beurt is jou beurt,* china.

12
Kudu Biltong

On a typical early morning follow-up, we were jaaging to a farm south-east of SWASPES where some terrs had caused shit during the night. It was just barely light, the sun wasn't up yet. I rode in my usual position at the back of the pack. To stay out of the dust, I kept a good 100 metres behind the oke ahead of me. I was half asleep and it was fucking freezing.

A bunch of kudus spooked by the noise jumped the fence and ran across the road in front of me. I saw five. I went between two huge males at 90. The one to my left still had his hind legs on the road and the next one's front legs were on the right edge. I didn't even see them till I was maybe 20 metres away. Just a snapshot of white road and dust, pale blue sky, dark bush, grey backs with white stripes and huge twisting horns, and I was through. I never let off the throttle or touched the brakes.

Kudu makes the best biltong you'll ever taste, but that would not have been a good way to get it.

After the new 'ProLink' XR 500s arrived, we spent a few weeks doing training with platoon 12. One of the first training rides south-east of Otavi, a few kays past the spot where I almost hit Kudus at 90 km/h.

13
Swapo Kos

Travelling by train through Windhoek one time with a berede platoon, I made the acquaintance of rifleman Lock, a shit-stirrer like Kruger. His platoon had shortly before been lectured by some high-ranking jam stealer that, if they didn't pull their socks up, they'd end up as 'SWAPO kos'. The insult resonated with Lock and he ran around for weeks afterwards torturing his mates, singing: '...*ek is tog SWAPO kos, naaa na na naaa na......*'

I once put a meal on the table for SWAPO during the farm incursion around Tsumeb and Otavi, but apparently they weren't hungry that day. We mostly did follow-up and tracking after contact, but also monitoring of waterholes, isolated shelters and possible sympathisers. Places the terrs would go to find food, water or shelter. I was sent with one of our sections that day to look for SWAPO activity, meaning spoor, at a specific farm dam. There had been a lot of movement in the area and the higher-ups' plan was denying the terrs resources in the dry bush.

The round, whitewashed dam had a trough around its base and the bush was cut back maybe 30 metres. The Buffel driver stopped on the tweespoor past the dam. I told the troops to wait, walked up the slight incline to the dam and began doing 360s around it, widening the radius every time I completed a full circle. I found nothing but cattle spoor and cow shit.

By this time I was a good if officially unqualified tracker, having spent many long days watching and learning from the okes with the ratel badges. Keen observation skills and correct techniques were half the battle. I swept all the way to the edge of the clearing and stood there staring at trees and rocks, debating whether to do a sweep through the bush. For some odd reason it was a weighty decision. I started towards the edge of the bush, then decided it would be pointless and turned back.

The troops roamed around the clearing and found nothing, so

we got back in the Buffel and left. Maybe fifteen minutes later, the German farmer who was the Commando officer in charge of the operation called on the radio and started kakking me out. He told me that some trackers had just swept the dam and found lots of fresh SWAPO spoor. He was unhappy and it turned into a rather heated discussion. I didn't appreciate his insinuations that we were incompetent. I later had words with him face to face and made it crystal that there had been no spoor when we swept the area, *verstehen sie?*

We were kept very busy and the incident was soon forgotten. It only dawned on me many years later that some SWAPO boffin terrs had walked up and got water out of the dam in the few minutes between me and the trackers sweeping around it. That meant they had lain in the bush and watched me through AK sights do 360s around that fucking dam.

I would have been in deep shit, had I walked into the bush by myself. For some reason I changed my mind at the last second. One of the better decisions I've made in my life.

SWASPES raison D'etre: Swapo alpha-sierra (AS) spoor on a farm south-east of Otavi during Ops Yahoo.

14
Frozen Moments

I can't tell you what I had for lunch yesterday but I can recall images and incidents, even conversations, from 1982 as clearly as if they happened 25 minutes instead of 25 years ago. When I close my eyes I see it like a slide show on a white screen. I can still picture every man of my platoon as clearly as if they are standing next to me now. I recognise individuals from silhouettes in my photos. I can still hear their voices. It's that clear.

While scanning my photos onto computer disc recently, I was delighted to discover that the resolution was good enough to enlarge them many times. I could see much detail that were lost on the small photos. With the objectivity of time gone by, I made some interesting observations.

Most of the photos are happy shots. We are frozen in the acts of wheelying, swimming, braaiing, and fucking around playing various dangerous games. One shot shows old MJG taking a *boskak* in Angola. Hilarious to the onlookers, but he was very unhappy about us documenting the occasion. He considered it an assault on his dignity. I also noticed how the bikes turned from brown to red as the months went by and the paint was scratched and rubbed off.

Captured too are the daily, deadly serious routines of setting up and sleeping in TBs, making radio reports from Angola, and fixing flats. Cooking our meagre dinners before moving a few kays away to sleep, hoping SWAPO would throw mortars where we ate instead of where we slept. Sweeping around cuca shops, looking for *alpha-sier-ra* spoor.

But many images exist only in my memory. Like the jam stealer Intelligence Captain who tried as hard as any terr ever did to kill me. He sported a green beret with a varkpan badge, shorts, knee length

Sweeping for spoor around a cuca shop in case a swap got careless and changed his uniform, but not his boots with their distinctive chevron patterns.

socks, and fucking *Grasshoppers*. A skinny, dark-haired, short-sighted *doos* with thick glasses that turned dark in sunlight.

The grinning Koevoet policeman throwing his FN Mag with ammo belt flapping behind it, over the high side of a Buffel and jumping after it in a contact.

The Buffel with its front end blown off that took the punch for my bike and I.

Four dead swaps, three with no heads, who got caught by Alouette gunships. Most of their kills were head shots, the gunners told us.

The look on a troop named Van Niekerk's face, tasked with taking the four smelly ex-comrades to Oshakati in a trailer behind a Buffel.

The tension on the faces of two dozen okes deep in Angola as we

25

silently push our bikes away from where we shut the engines down for the night's TB.

The images appear in no particular order, at random times. I wish that I could print some of them. Like the smiling face of the muscular parabat Captain standing in that dusty Otavi street four hours before he was killed by a terr hand grenade. He was wearing his maroon beret, his long sleeves were rolled up just above his elbows, and his brown pants were faded yellow.

Or the ridiculous sight of our tracker drinking buddy who was shot in the groin and bled to death, parading around in skimpy red panties and bra during a major pissup at SWASPES one evening. I always wondered where the fuck he got those. Trophies of some wild night back in SA, was all he ever said about it.

Or the mixture of shock and amusement on old Van's face when he leisurely pushed down on the kickstart trying to find top dead centre with his bike in gear, on the sidestand, and it fired up and took off.

Innocuous sensory inputs trigger memories that constantly surprise me. Seeing a tall palm tree, a Honda XR 500 or a full moon. Hearing certain songs or the metallic whine of an Alouette flying by, a braying donkey, or truck engine noise of a certain pitch. Strongest of all, the smell of diesel on a hot day, an embedded smoke smell only found in rural third world countries, or the sour pong of sweat-soaked clothes.

I enjoy my private slide show. It makes me smile or laugh out loud, and I hope it never fades. So far so good.

15
The IQ of Swine

We chased a group of swaps around an area south-east of Otavi for more than a week. On one of those mornings, quite by chance, we saw three cheetahs the local farmers had been hunting and killed two of them. We saw kudu and warthogs every day. Breeding season must have ended shortly before, as many swine families with brown, hairy little ones were out and about.

Few animals are as amusing as a warthog. They're so ugly you can't help but like them. Flying around corners or jaaging through the bush, we sometimes surprised them and they'd run for their lives. A running warthog is a comical sight. Their tails go straight up in the air and their short fat legs scissor back and forth at blinding speed.

When we spooked them on the road, they ran along the fence looking for a gap to duck into the bush. We'd slow down and follow, laughing our arses off as they upshifted until their legs were a grey blur, trying to evade the big noisy predator chasing them. Most of the bigger ones I saw were three-speeders. Those with steadier nerves just knelt there arse-up on their elbows digging for food, and gave us a look that said '...*piss off, I'm trying to eat lunch here*'.

We saw dozens every day for more than a week. But no SWAPO, except *alpha-sierra* spoor. On the last day, we schemed a warthog would make a nice platoon braai back at Spes. We got permission from the farmer whose land we were on to shoot one, and drove around for hours looking for a nice-sized pig.

Somehow they knew we wanted to shoot them. We didn't see one fucking pig all day. And people think horses are intelligent.

Our improvised track at SWASPES named 'Back of the Stables' after the MX track Back of the Moon, north of Pretoria in the Transvaal.

Puma at SWASPES, April/May 1982

16
A Swinging Swapo Safari

The bush war was fun and games at times.

At least, bike squad's part of it was. During Ops Yahoo we were rewarded with a trip to the Etosha Game Reserve, a holiday of sorts. Except we had to earn it by roaming around the isolated western part for a week, looking for SWAPO spoor on the cutlines and roads that traverse the park. The western half of Etosha is closed to the general public to this day. Only Park Service personnel and professional tour operators are allowed there. In 1982 it was limited to Park Service and the army.

We rode our bikes from Otavi to Okaukeujo where we transferred to Buffels. It would be dangerous to ride bikes amongst the animals, they told us. Ja well, no fine. Late in the afternoon we jaaged through Okaukeujo gate in a cloud of white dust, past two tour buses full of pale Europeans 'on safari'. I was much happier carrying an R4 than a Canon under the circumstances, and thought they were either very brave or very foolish. Judging by the big eyes staring at us from the bus windows, some of them thought so too.

We had three 101 Bn Ovambo trackers assigned to us. They took turns sitting on the front of the Buffels, looking for spoor as we slowly drove along the roads and cutlines. We were bored shitless, but the grumbling stopped once we started seeing thousands of zebra, oryx, springbok, giraffe and other antelope. We stopped to watch a herd of more than 60 elephants wander by, and stared fascinated as giraffes spooked by the Buffel noise ran alongside the road in surreal slow motion, their long necks and legs waving back and forth slowly like tall grass in the wind. We were used to seeing the blur of running warthog legs in the bush around Otavi and were thrilled by their graceful flight.

Flat terrain with patchy thorn bush and grass gave way to shimmering mirages on the white surfaces of Etosha Pan and the smaller ones to the west. The pans may look like solid salt, but are actually

Giraffes in Etosha, crossing one of the roads we swept for SWAPO spoor.

dried white mud. We crossed the smaller ones a few times and the drivers loved putting foot on the good surfaces. That made us very unhappy about not being on our bikes.

Crossing the first pan, the oke driving the lead Buffel got a bit of a surprise when he jaaged up to the edge of it. As did we in the back. The bright sunlight reflecting off the white landscape fooled his depth perception, and he didn't realise until the Buffel was airborne that the surrounding surface was higher than the pan. He hit the edge balls out and jumped that Buffel a good fifteen metres. Standing up in the back, we were tossed around violently. A seven-ton mineproof vehicle doing 80 doesn't exactly glide in for a landing.

Laughing and yelling, we looked back and were very impressed with the gap between the edge of the pan and where the Buffel tracks started. The harsh reality that, if the front wheels of the top-heavy and unstable Buffel had been turned even *slightly* there would have been some more unusual markings as well as broken bodies on the ground, didn't bother us.

Carefully approaching the edge of another pan later that day, we found that slower wasn't necessarily better. The dry white surface concealed thick mud

Getting a Buffel unstuck in Etosha.

underneath and the Buffel got stuck up to its axles. The driver's skill and the excellent off-road capability of the Buffel's Unimog chassis got it unstuck. By rocking back and forth he managed to reverse out of the mud back to solid ground.

Nights were spent sleeping under the stars in the denser bush further west. I organised a guard system with two okes awake at all times, which gave us all a good night's sleep almost every night. We made big fires in the evenings, figuring the swaps wouldn't be foolish enough to attack a full platoon.

Just as well we didn't know then what I do now, that we were looking for three groups of 40 each of SWAPO's specially trained *Volcano* terrs. It may have put a damper on our camping trip.

One of those evenings around the fire, we accidentally discovered an amusing new game. A section leader changed the battery pack in an A53 radio and chucked the old one in the fire for lack of a better idea. Those 9V battery packs were slightly taller and skinnier than a beer can, with six 1,5 V cells in series held together with wax. Two or three minutes later the battery pack exploded with one *moerofa* bang, powerful enough to send burning logs flying all over the fucking place. Bit of a conversation killer, it was. Two Ovambo trackers did synchronised backflips off the ammo crates they were sitting on. A few okes who didn't see the battery pack thrown into the fire bekakked themselves, grabbed their R4s and ran around looking for something to shoot. I had watched the whole circus unfold and rolled around laughing my arse off.

Game on. The next few evenings, fires exploded randomly and scared the shit out of whomever wasn't paying attention. I pulled the same stunt in Ovamboland later with the same hilarious results. After crawling along roads and cutlines for a few fruitless days, our SWAPO safari ended and we returned to SWASPES and the intense activity of Ops Yahoo.

I have not been back to Etosha since. Photos my father sent me of his recent Namibia trip instantly brought back memories of the relaxing few days we spent there. We never realised how privileged we were, roaming around one of Africa's great game parks with total freedom and authority.

Southern Africa is very different now than in 1982. I will return to the blinding white pans and endless flat horizon one of these days. I probably won't blow up any fires, but this time I will carry a Canon which is just fine by me.

17
Musikladen

In the late seventies, TV in South Africa was still in its infancy and a show called *Pop Shop* was daring and controversial. Thirty minutes, once a week, of music videos and live bands presented by an oke named David Gresham, was considered cutting edge and a bit rebellious in those days. He was a smooth talking, sharp dressing, with it, cool cat. His show was a glimpse of the outside world the youth of the isolated country had never seen before.

At SWASPES we got hold of a German video called *Musikladen*. Meaning, 'music shop'. The hour-long tape was similar to the TV show in SA except for its distinct European flavour. The Germans love their naked *punda* and Musikladen had plenty of it. One song had a bunch of genetically gifted girls in fishnet mini dresses and no underwear bumping and grinding on stage in full-colour detail. We wore that fucking tape out.

After closing the Otavi Hotel bar on Saturday nights, we normally woke up the barman at the stuffy and boring officers' club and bought a few bottles of Red Heart Rum and Coke. Back at the school we lived in, the booze, the snooker table and singing along with Musikladen, over and over, provided hours of noisy and destructive entertainment. One of those nights was the first of two occasions I have ever seen broken snooker balls.

We played the song with the dancing *punda* more than 50 times one evening. I know, because we counted. Our tracker friend Dave got nostalgic and slipped into red lacy bra and panties he had *slikked* from some chick in SA whom he knew in the biblical way. They must have come off a shapely woman, judging by the sizes, and we were appropriately envious. But we really didn't need to see that much of Dave. He didn't have the legs or the tits for it. We sprayed him red with mercurochrome and made him go change.

I wish I can remember the name of that song. It was a catchy tune popular in Europe and SA at the time, I know I'll recognise it if

I hear it again. I also wish I had taken a photo of Dave that evening, even if he was giving lingerie a bad name. Some months later he was running on spoor and unexpectedly caught up with the terrs he was tracking. He was shot in the groin and bled out before a medic could get to him.

18
Things That Go Bang In The Night

Contrary to some revisionist rubbish written lately, we did lots of night ops, even in the dense bush around Otavi and Tsumeb in 1982. I have since read accounts of the conventional battles in Angola from 1986-88. Those okes did little but hide from MiGs in daytime. Much of their movement and fighting was done at night.

My platoon was very active before sunrise during the farm incursions. By the time the sun came up we'd typically be out in the bush somewhere, chasing spoor after farmers reported terrs showing up at farm workers' houses during the night and causing shit. I don't recall them ever killing any workers, they'd just push them around some. I always suspected that they were received with open arms in some cases.

As part of that same operation, we once spent all night riding up and down the railway line between Otavi and Tsumeb. We had to make as much noise as possible to drive the swaps away from the rail line, towards other units lying in ambush. Or so the theory went. We thought there was a very distinct possibility of getting ambushed ourselves.

To ensure SWAPO saw and heard us, we loaded up on flares, 40mm rounds and as much 5,56 tracer as we could carry. We usually only loaded two or three tracers near the bottom of each magazine, but some okes had seven magazines full of it that night. I grabbed a box of thunderflashes just for a laugh. They were basically giant crackers capable of blowing your hand off if you got careless with them.

The platoon gathered at the railway crossing on the west side of Otavi just before sunset, and we set off at a slow pace after dark. The loot and I were in the lead side by side, with the rest of the bikes strung out behind us. We were tense. The service road we rode on was a tweespoor tunnel through tall grass and dense bush, and the puny lights on the XRs reached out only 20 metres or so. The bush alongside was pitch dark, two or three well-trained terrs lying in ambush would have slaughtered us. After we'd crawled along for an hour or so and hadn't been mowed down, we schemed we would survive the night and resorted to fucking around to make it go by quicker.

I soon got bored and dug a thunderflash out of my webbing. By holding it with my throttle hand, I found I could run the striker over the fuse with my left, like lighting a match. The fuse burned for fifteen seconds. I dropped it next to my front wheel, started counting, and told the loot to look around. As we did, the thunderflash went off with one *moerofa* bang right in front of rifleman Franken three bikes behind. Next thing we saw was a headlight beam appearing out of a wall of dust, swinging from side to side as he battled to stay upright. I almost fell off my bike I laughed so hard, but the okes close to the explosion were not all that amused.

The game was on. For the next hour I lit thunderflashes and dropped them on the tweespoor, and the riders behind me weaved and sped up or slowed down to dodge the explosions. It turned into a cat-and-mouse game, as I dropped the thunderflashes after letting

Platoon 11 next to the railway line in Otavi after riding up and down the line to Tsumeb all night. Herbert the German loot at left and me at right, very amused about the night's fun with thunderflashes. The rider right behind me was not.

the fuses burn for varying counts. Nobody was sleepy anymore. One of my less intelligent moves of the night was holding on to one for so long that it went off right behind my bike, two seconds after I dropped it.

I probably played the game beyond the point where it was funny. One oke named Michael was very perturbed with me when daylight came and we were given something else to do. He was very unhappy about his front fender. It was cracked halfway through with many rectangular holes from pebbles blasted through it by an exploding thunderflash. In the photo taken at the Otavi railway crossing that morning he is right behind me, visibly annoyed.

During the night we stopped at several predetermined spots and fired off the ammo we carried. It was quite a display and a good tension reliever, but we had our doubts about the objective even as we blazed away at the night sky and dark bush. Two round trips of 120 kays each, with a break in between during which some okes tried to catch a nap, used up the night.

When the sun came up we expected to get some sleep, but were sent in pursuit of something else after breakfast. After another long day we finally got a decent night's sleep, having been awake and mostly riding for 36 hours. I was exhausted but having the time of my life. Everything seemed funny. I felt drunk without the impairment or hangover, which was quite pleasant. Another interesting new experience courtesy of *PW's Poes Plaas*.

After jaaging around again all the next day, we were sent on

On a bigwheel XR 500 during Ops Yahoo, April 1982.

another nocturnal caper. Somebody hatched the clever idea of laying an ambush along a donga east of Otavi where the road makes an S-turn through the hills into the valley to Grootfontein. One section ambushed the donga and the other went up the hill to a spot where they had a spectacular view of fuckall and the stars. We had to walk in from some distance away, ducking through the thorny bush in the dark. It was cold and windy and the okes up on the hill were really miserable. We hadn't been back to Spes all day and nobody had bush jackets or jerseys. It was a very kak night.

Around 04:00 we heard something coming down the donga. Adrenaline flowed and suddenly nobody was tired. When it was right in front of us, two okes threw illumination grenades and everybody aimed at the noise with fingers on triggers. After a few tense moments we realised it was some kind of large animal. We never did see it clearly. Having given away our position, the ambush was over. The section on the hill was called down and we walked back to the road. I don't remember how we got back to Spes, but by the time we did it was daylight and we got back on the bikes for another day of blurry activity. I noticed the same feeling of detached amusement as two days before and realised that it was a side-effect of sleep deprivation.

We had been up for 36 hours, slept 8, then up for 36 hours again. I later read that depression is sometimes treated with controlled sleep deprivation. That's one reason spec infantry was such a jolly place, I think.

The local commando Major who was in charge of the show, finally warmed up to us after another sneaky night operation. We had been on less than friendly terms ever since the farm dam episode. He thought we were Joburg wankers and me particularly so, a city slicker who didn't know his arse from a chevron spoor. He was openly hostile when he briefed me on a two-man job one evening. With one sharp troop as backup, I was sent to a farm worker's house to observe him through night vision binocs. Based on information he received, the Major suspected the worker was providing the last remaining swaps in the area with food and shelter.

The house was a ways from the tar road, on the north side of the valley east of Otavi. It was a quiet evening and the distinctive noise of two Buffels could be heard a long way. To avoid alerting anybody in the area that the *boere* were up to something, the one we rode in slowed down but didn't stop as we jumped off. Both vehicles continued on towards Grootfontein.

It was a piece of cake. We sneaked up so close on the farm worker that we hardly needed the night vision. We lay within spitting distance while he made a fire, cooked dinner, stared into said fire for a while and eventually went to bed. The Buffels picked us up again and soon after I was debriefed by the Major. He asked me how close we'd been. When I told him less than fifteen metres and gave a detailed account of what we'd seen, he quite obviously changed his opinion of us. After that night we got along famously. The Major drove an army Mazda bakkie and his German pronunciation 'Mats-da' always amused me.

After many weeks of intense effort, the terrs we were chasing had been whittled down to one or two and the higher-ups were thinking up all kinds of schemes to catch them. The last one was known to be in the narrow valley that runs from east of Otavi almost to Grootfontein and we did many small raids on various locations, trying to snag him.

One full moon night the platoon was sent to a shack in the bush where this elusive swap was thought to be. A parabat Captain had just been killed under identical circumstances, and we took the seemingly easy outing very seriously. We were dropped off on the tar road several kays away and sneaked along a tweespoor leading to the shack, set in a clearing in the dense bush.

Every full moon night I still see that scene: two lines of dark silhouettes moving silently along the sides of the sandy track, through patches of moonlight between the black trees. The freezing cold and bright moonlight somehow made the corrugated shack in that small clearing seem very threatening. We set up in an L on two sides of it and someone kicked the door in. It was empty. After sweeping the area for spoor with torches we walked back to the road, talking and laughing. Everybody was quiet on the miserably cold ride back to Spes in the open Buffels.

The last big night operation we took part in was a search of the mineworkers' hostel at Kombat mine by Koevoet. A convoy of Buffels and Casspirs charged up at high speed and surrounded the large two-storey building. A few okes chose to go in with Koevoet but instinct told me to stay in the Buffel. Another good decision, it turned out. They came back green around the gills and later told many disgusting stories about what they saw. A roaring prostitution trade went on in the hostel, and they learned a bit more about hygiene and how the other half lives than they had bargained for.

The operation that had lasted several months ended in somewhat of an anti-climax. The last remaining swap that we and others had chased so long and hard, was captured by a farmer and his son out hunting one day. They caught him at a spot on the hillside above the local abattoir where I once sat for hours, breathing through my mouth and waiting in vain for him to pitch up. Inconsiderate bastard.

19
Breadrolls And Linda Lovelace

The junior officer and NCO sleeping quarters at SWASPES were part of the vacant high school in Otavi but the mess was on base, two kays east of town. Once SWAPO's farm invasion started winding down, weekends were very *rustig* provided you gyppo'd out of being duty officer or NCO.

We soon established a routine. We'd sleep late on Saturday morning after a minor pissup Friday evening, then wander into town looking for amusement. At that point the first priority was food, and going all the way out to Spes for breakfast was a hassle. Not to mention, it was usually closer to tea time when we surfaced. Rum hangovers are rough, even at 20.

On the corner a block from the school was a little bakery owned by a traditional German baker. The flat-roofed brown brick building sat diagonally across the corner and the big display windows slanted outwards. I forget the baker's name. He only spoke German and looked rather intimidating. He had a big mop of curly black hair under a white baker's hat matching his white coat, a black moustache and goatee, wild eyebrows, and a hook nose. Fiery dark eyes completed the picture. Dressed in red he would have looked like a Shakespearean Lucifer, but he was a nice oke once you got to know him. We had seen him baking at 04:00 many times during the early morning follow-ups of the previous months. The smell coming from the little bakery that time of the day was wonderful.

We went into his bakery at every opportunity for more than his bread. He had a stunningly beautiful daughter named Jutta. She had long black hair, spoke English and seemed fairly shy when four or five testosterone-charged twenty-year-olds walked in, but it could have been plain fear. She drove us crazy but we were wary of her father and treated her with the utmost respect.

The baker made *Brötchen* every Saturday morning. Platoon 11's loot was German and he explained that it meant 'little bread', bread rolls. Our favourite was *Brötchen mit Kudu*. The rolls were still warm, with fresh butter and thin slices of smoked kudu meat it made the best breakfast I've had to this day. We each stuffed three or four down our necks and put a serious dent in the morning's batch, but he didn't mind one bit.

If Jutta wasn't around, we wandered down the street after decimating the town's weekend *Brötchen* supply. Next on the agenda was intellectual stimulation, so next stop was the BP station in the middle of town to buy a magazine or two. Only Bike SA or some other motorised publication held our interest in any way and the petrol station shop wasn't exactly a library. Therefore we usually arrived at our final destination by 11:00, intellectually deprived but ready for the next activity on the schedule. The main event, really.

Around a corner from the petrol station was the Otavi Hotel. More specifically, the bar in the Otavi Hotel. We became friends with the German owner and his son, who was our age. I can still see them behind that bar counter. The old man wore thick glasses with a light-coloured frame, a pale blue golf shirt with stripes across the chest, and shorts. He had thinning gray-blonde hair, sunburnt skin and a permanent smile on his face. His son was average height and muscular, with olive skin and thick dark hair over his ears. He wore a bright blue T-shirt and denim jeans and he was leaning on the bar counter talking to me.

The son was not required to do military service because he wasn't a South African citizen. While talking drunken nonsense with him, I remember sensing the divide between my world and his. I felt trapped and old compared to him. We spent several happy, boisterous Saturdays in that pub. A primitive early-80s video game involving racing of some kind provided hours of noisy amusement. We became expert at it and wrote our names and record scores on the wall above the machine. I wonder if they're still there.

The hotel owner liked us. We spent lots of money in his pub and invariably ate lunch and dinner there too. After he got to know us, he closed early one night and invited half a dozen of us to watch a movie in the lounge next to the bar. His son sat to our right, on the windowsill in the thick sandstone wall, and kept looking outside. He was fidgeting like a whore in church and he was making me nervous.

Once the film started, I knew why. It was *Deep Throat*.

Now remember, this was 1982 and we were essentially in South Africa, where many things were banned as *Die Rooi Gevaar*. According to our moral guardians, porn films were distributed by the communists for the sole purpose of exterminating the *boere*. I remember feeling a mixture of shock, envy, amusement and enlightenment. I had seen headless and otherwise mutilated swaps but until then, not the demoralising marxist images on that screen.

At least not ten times larger than life in close-up, full colour detail.

I wasn't the only one in that room having his horizons expanded. You could hear a pin drop in between the moans and inane dialogue. Most of the other okes were staring open-mouthed too. We certainly did learn a lot about life that year. I still don't know whether I should blame or thank the army for it.

Braai at the school where we were quartered at SWASPES. The three men at left and the one at right looking at the camera were veterinarians. In centre with white T-shirt is Platoon 12 commander Earl, then Herbert the Platoon 11 pelbev, a berede pelbev named Buks, a veterinarian staff corporal, and me in red at right.

20
Civvy Kak

A surprising twist in the tale was the clashes we had with white civilians in Otavi and Tsumeb during and after Ops Yahoo. We didn't consider ourselves crusaders or keepers of the flame, but we certainly didn't expect hostility from the people we were trying to protect. A few farmers were upset about the fences my platoon, 61 Mech's Ratels, and Koevoet had cut or simply driven through while chasing terrs. Despite getting reimbursed by the army, some remained hostile and it spilled over at times.

On the western edge of Otavi was the white, fort-like German Club. One of those joints with a 'membership fee' high enough to keep the riff-raff out, that you still find all over Africa today. Most of the white civilians in town spoke German as first or second language. Since platoon 11's loot was a German from Otjiwarongo just down the road, we were accepted and allowed to drink in their pub.

One of the social events the Club organised was the Otavi Rally. It was a race around the local area, not about speed as much as following directions to find objects hidden at specific locations. A motorised treasure hunt of sorts, with a time element to keep things lively. The new loot and I were dispatched with a section from platoon 12 to represent the SADF. It was a pleasant distraction and we took part with great enthusiasm.

Initially we followed the route map and collected the information needed at each checkpoint, but soon the need for speed overcame us and we jaaged the route balls out. I knew the roads well from chasing SWAPO around the area over the previous months, and we beat everybody back to the Club. However, without the completed navigation sheet, we were disqualified. No worries. We got our dose of twisting throttles and drinking beer and nobody crashed. A perfect day.

A section from Platoon 12 was dispatched to represent the SADF in the Otavi Rally, a motorised treasure hunt organised by the local community. Here the racers are gathered at the German Club before the event. L to R: Le Roux, Reeves, Welsh, Beudeker, De Jager, Douglas, Hagen. We were disqualified but we got to jaag around and drink beer, which made it a perfect day.

A few nights later I went back to the German Club with the new loot. He was almost two metres tall like me, 19 to my 20 years old, with a temper even shorter than mine. A local in white shorts and a pink golf shirt with straw-like yellow hair, a red sunburnt face and round monkey eyes started chirping us. I had the impression that his fence had been flattened and that he was a grudge bearing type of sausage bender. He made it clear we weren't welcome. Things escalated quickly. Insults, chairs and tables and punches flew before we left for the friendlier environment of the Otavi Hotel.

While Ops Yahoo was still in progress, a berede loot came up with the bright idea of going to some big jol at the Tsumeb show grounds one Saturday evening. I forget his name. He was a stocky and aggressive blonde oke just returned from a nine-month stint in Ovamboland, and he was in the mood for action. Only problem was, civilians were prohibited from travelling at night to prevent ambushes and the army would most likely turn down a request for a Buffel or Ratel to escort us. After some debate we solved the problem by concluding that we weren't civvies, no matter what we wore. Four of us piled into the German loot's red VW Golf and jaaged the 60 kays to Tsumeb with cocked R4s sticking out the windows. SWAPO would have gotten one *fuck* of a surprise had they ambushed that civvy car that night.

Arriving safely, we locked the R4s in the boot and ventured forth in search of punda. It didn't last long. Tsumeb is a mining town, and soon a bunch of local *breekers* wanted to fuck us up. Vastly outnumbered, the four of us were surrounded in semi-darkness between two big tents and threatened with extinction.

But the miners had picked the wrong okes to fuck with. We were spending our days trying to kill people and were in an aggressive state of mind, to put it mildly. The blonde loot and I slipped away and fetched the guns out of the car. Next thing the *mugus* knew, they were staring down the barrels of cocked R4s with safeties on 'R' for Rock 'n Roll. The flashguards up their noses made their bravado evaporate like *mieliepap* farts in the wind. After they retreated behind a barrage of insults and threats we lost interest. We knew they'd be back with half the town and that blood would flow. We reluctantly ran the gauntlet back to Otavi and our normal weekend routine of closing the Otavi Hotel bar, followed by Red Heart and Coke, snooker and *Musikladen* in our quarters.

After the Otavi Rally. L to R: Maree, Hagen, Douglas.

With hindsight I prefer to think we were only scaring them off that night, but remembering our violent tendencies, I'm not sure at all. I still haven't lost the habit of reacting violently to perceived threats and it's difficult to control sometimes. Not a very socially redeeming trait in some circles, but Machiavelli was right.

It tends to solve problems before they even start.

Platoon 11 on farm protection duty west of Otavi, April 1982.

Another braai before another rough morning. With our mate the SWASPES signals NCO.

21
Papio Ursinus

A wide variety of pets were kept in every base on the border. Dogs, cats, donkeys, monkeys, birds, goats, you name it. The *recce* base had a tame-ish lion. They were a distraction from the harsh conditions and a reminder of home.

During Ops Yahoo, platoon 11 acquired a baby baboon from one of the berede platoons. They, in turn, had inherited it from the cooks at SWASPES. The okes were very chuffed with their exotic and cute, if ugly, pet. Nobody questioned why the animal was passed around camp like the latest SCOPE. They soon found out.

In two weeks, four different okes in the platoon played daddy to the little creature. He was 20 centimetres tall and weighed only a kilo or two. But gramme for gramme that little bastard packed more destructive power than a fucking cheese mine. By the time he attacked Kruger in the Ratel he'd been shuttled around the platoon like a red-headed orphan.

After some platoon errand one day, the troops called me to one of their tents. They were extremely unhappy. They had left the little baboon in the tent with the side flaps up so he wouldn't escape. It was one of those 32-foot tents that housed ten troops, and it looked like a bomb had gone off inside. The baboon had gone on a rampage, clearly annoyed about being confined for two hours.

He destroyed everything he could get his little hands on. *Kaste* were open, clothes strewn all over the tent, laundry soap spread around like snow, toilet kits torn open, toothpaste squeezed out on beds and clothing. Every box and plastic bag in the tent was ripped open. Books and magazines were shredded, and he'd shat on several beds. I laughed so hard I couldn't speak. The troops were not amused.

That was the final straw. The next day they gave him to the dog handlers, who were used to taking care of animals. That little baboon was like a psychotic baby. I'm willing to bet that most of the okes who lived in that tent at Spes never had children later in life.

22
Asterix

When we got to Ogongo, the HQ of 52 Bn, I renewed an old friendship of sorts. Turns out the RSM was Sergeant-major Kotze, who had been my CSM in Alpha Company at Oudtshoorn in 1980. He was nicknamed 'Asterix' because he looked just like the cartoon character. He was short and lean with pale blue eyes and a huge walrus moustache. Had his short brown hair been long and blonde, he'd have been the spitting image of the little Gaul from 52 BC.

Asterix was the single most amusing character I encountered in the army. He had a sharp wit, a foul mouth, and a slightly high-pitched voice that carried a very long way. He was also rather relaxed for a Sergeant-major, without the excessive *houding* and nasty disposition many of them adopted. His salute was a casual wave somewhere above his eyebrows, and he treated the NSM loots like troops. Yelled at them and told them what to do. Funny thing was, they jumped and complied.

But Asterix was far from slack. He just didn't partake in a lot of the mindfuck that went on. He was respected and liked by those both above and below him.

Every senior NCO had his own vocabulary and Asterix was no exception. He used the nonsensical but rhyming phrase, '...troepeeeee...julle moenie my moer koer nie' in a few variations regularly. I remember him yelling at an MP we had in the platoon in 1980 one morning: '...kopperaal Raubenheiiimeeer!!! ...maak die plante nat, dis so droog soos 'n non se kont!' The entire company invariably ended up in stitches whenever he addressed us.

He remembered me too, I don't know why. Early each morning of the few days we spent in Ogongo, everybody had to *tree aan* to be assigned chores for the day. The first assignment was always '...kokke kakke en klerke fokof...gaan doen julle ding.'

Next he yelled at me. I played the bike maintenance card from day one. We liked to keep outsiders in awe of our assumed mechanical prowess, as it created leverage in potential rondfok situations. At Ogongo, it worked well. By the third day Asterix didn't even ask, he just told us to piss off and go do *onderhoud* while everybody else picked up cigarette butts or dug shit pits. I caught his eye one time, he all but winked at me. We exchanged knowing looks and I gave him a slight smile of thanks. He had it all figured out.

Asterix was a good oke. He and a one-armed Rhodesian Sergeant-major at Spes were the only two people of that rank that I ever actually liked.

The loot and me at Ogongo, the HQ of 52 Bn. Doing 'onderhoud' while others swept the sand and picked up leaves.

23
Platoon 10

Bike platoon 10 was formed from the July 1980 National Service intake and went to Ovamboland in May 1981, four months after my arrival at Berede. My friend John from Durban, who had been in my platoon at Infantry School in 1980, was one of the four corporals assigned to bike squad before I porky-pied my way into the small unit. He was attached to platoon 10, and the other new corporals were scattered around bike squad in various capacities ranging from workshop manager to training the new intake of recruits. Bike squad already had five corporals on staff and its personnel roster became rather junior NCO heavy. My unexpected appearance, which everybody assumed was usual army left hand-right hand confusion, and didn't question – they could not have dreamed how accurate that assumption was, albeit not quite the way they imagined – made me a *spaar piel in die hoer huis*. Consequently, I floated around Berede trying to stay under the RSM's radar for the entire year.

My *gatvol* meter pegged around May and I requested a transfer to 32 Bn, but the papers were torn up by a vindictive cunt of a Major appropriately called 'Ratsy'. I was furious at the time, but things happen for a reason. I would not have these stories to tell or own fifteen motorcycles, had Ratsy not stuck his pointy nose into my business.

The day in September 1981 when it snowed in Johannesburg was brutally cold and windy on the open plains around Potch. By that time platoons 11 and 12 were on their way to Ovamboland, and bike squad was composed almost entirely of *spaar piel* corporals. We spent the day curled up under blankets in the troop bungalow furthest from the offices and high-traffic area. Nobody bothered us all day.

John was a member of the Durban KDX 200 off-road mafia and a much better rider than me, seeing I had only started riding in January. We had some memorable adventures including an epic trip to Durban over Easter weekend. I disassembled my Suzuki PE 400 – my first bike, a nasty brute of a machine which I had no business riding with my minimal experience – and crammed it into the backseat of his Mazda B1300, a little shitbox car smaller than a VW Beetle.

After the 10-hour drive from Potch in long weekend traffic and several hours reassembling the PE, we went for one ride in the hills and spent the rest of the weekend making the rounds of all his Durban jols. We went to the races at Roy Hesketh circuit with John's father and his mates on Saturday 18 April 1981, and got *vrot* over fish and chips at an English pub on the way home. After dropping the *ballies* off, John and I proceeded to a hotel in the hills where a friend's band was playing. After arriving at the venue, we both passed out in the car and woke up when the lights came on and the band packed up. So as not to appear unsociable, we went to an after party at another hotel with them. Many years later John's sister told me that their circle of friends still spoke of the graceful swan dive I did into the river next to that hotel, which was news to me. I remembered the river bank, but not my acrobatics or the fact that it was autumn and the river was dry.

Shortly after Easter John and platoon 10 left Potch, destination *grens*. They had a slow start as bikes were scarce and the army confused. Late in the year he crashed and broke both arms and was casevaced to SA, but not before he had accumulated some excellent adventures. John now lives in Australia, where he owns a petrolhead café and bar in the mountains outside Brisbane and more bikes than me. I visit him as often as possible, courtesy of my very mobile job.

The next two stories are from his time with platoon 10, in his own words.

24
Kurpene Swapo

Platoon 10 deployed to Ogongo in autumn 1981. At that time there was a waiting period for XR 500s, so we were The Bike Squad in the operational area with no bikes. The camp commander must have been gatvol of us hanging around and doing the occasional patrol, and somehow 30 bicycles were organised from somewhere. They were well used, not fancy, those kak type of bicycles that you saw the locals casually pedalling along the roads and across the shonas at their slow pace. Heavy bikes, with fat tyres and metal brake rods and so on.

We were given orders to patrol a certain area on our 'bikes'. The plan was, we would be taken out in Buffels and dropped off under cover of darkness. The bikes were loaded onto a Buffel and we all got in. The transport travelled the roads and then veered off into the bush so we could disembark in the dark. Offloading was a bit of a mess, trying to untangle the bikes and work out which was whose in the pitch dark.

That done, we started pedalling to get some distance between us and the drop-off point. As platoon sergeant, I rode at the back. When I began pedalling I thought that it was going to be a bugger of a patrol. Even on the flat I was battling to pedal, and I was losing ground to the rest of the platoon. It seemed like we pedalled all night. I was huffing and puffing to keep up, but the rest of the guys didn't seem to be taking as much strain as I was. We set up our TB and I went out like a light.

The next morning, after the usual breakfast and *boskak*, I inspected my bike and found that one down tube to the back axle had been squashed against the wheel. It was like a permanent handbrake. It must have happened when the bikes were all loaded onto the Buffel like scrap metal. We bent the pipe back using an R4 for leverage, and I got on to give it a try. Compared to the dead weight I pedalled the night before it felt like I was on a rocket.

For the rest of that patrol, all went well except for the landmine.

We moved through the area on the bicycles, doing the normal stopping off at kraals and asking the locals '...*kurpene SWAPO?*' (where is SWAPO?) as if they were going to straight out tell us. We also said a fair amount of '...*heela!*' (come here) with the same result.

We took rest periods under the trees on the edges of the shonas and ate our ratpacks. Depending on a person's taste, there was always a bit of swapping going on. The milkshakes went first along with the cheese and biscuits, but bully beef and peas were usually last to go. We caught donkeys that were stupid enough to get too close and rode them until they bucked the rider off. If they didn't want to run they were given a bit of incentive by inserting a stick in the area underneath its tail. It's funny how they became more active after that.

One evening when it was almost completely dark, we approached the area where we wanted to TB. I noticed a small mound of sand just inside the bush that looked suspicious, maybe a mine. We couldn't figure out what it was, so I halted the patrol and went to take a closer look. I had to use my hands because we didn't have any *soek-steek-stokke*. I wiped the sand away slowly and gently so as not to get blown sky high by the landmine underneath. As I got to the bottom of the pile, the sand was wet and then I felt something soft. Not sure what it was, I instinctively smelled my hand and nearly hurled on the spot.

I worked out that a local had crapped there and kindly covered it up with sand, like a cat. I don't know if he had covered it in a nice pile on purpose to make it look out of place, but if he did, it worked. I had shit all over my hands. The other okes had to empty their water bottles over my hands to get it all off. As much as I tried to wipe my hand in the sand and use my precious water to wash it off, I was sure I could smell that *pap* shit smell for weeks.

I swear, if I had seen a PB after that I would have really fucked him up.

25
Holiday In Angolia

(To the tune of The Dead Kennedys' *Holiday in Cambodia*)

During our time at Ogongo, while waiting for our bikes, we were sent to provide protection for a trip into Angola during Operation Protea. We were part of a big convoy of Buffels and Samil 100 Kwêvoëls. As far as I remember, there was a journalist with us and he was constantly taking photos. I had the impression that we were not part of the planning and were just hired guns along for the ride.

The first few days were uneventful and we travelled through towns in southern Angola like Cuamato and Xangongo, as well as smaller villages whose names I don't remember and probably didn't know then. There were visible signs of fighting in the towns with bullet marks in the buildings, craters in the roads and damaged vehicles. There was also a shortage of local population with a lot of deserted buildings.

One day when we stopped for lunch in a dense area, I noticed that one of the Buffels had a particularly flat tyre. They were intentionally let down a bit, as the sand was quite thick and it was easier for the vehicles to get through it with slightly deflated tyres. I told the driver, he acknowledged and I thought nothing more of it. After lunch we climbed into the vehicles and carried on the journey through the relatively thick bush. Some of the Buffel drivers were like cowboys and knew only one speed, full throttle.

Driving through the bush we suddenly came across this scene that looked like a Buffel had hit a mine. It was on its side and there were soldiers lying everywhere. As we were first on the scene, we started providing first aid to the wounded. One guy was visibly finished as his head was a distorted shape. I suspected the Buffel had rolled over him so I left him as there was not much we could do for him. I moved on to a guy who was breathing, but emitting obvious signs of distress and battling to breathe. I told one of our troops to get the medic kit and tried to get a butterfly drip into him. Not being a medic or knowing much about medical things it seemed like the best thing to do. I was battling to find a vein which was obviously due to his low and lowering blood pressure.

He was a smallish guy with red hair, funny how sometimes you remember things and how sometimes you can't forget them. I didn't succeed in getting the drip in before he died.

The rest of the day is a bit vague. I remember the choppers came in and the dead and wounded were taken out. I can't remember what happened to that Buffel, but after that we strapped ourselves in. In my opinion that Buffel driver caused those deaths as he was aware of the flat tyre, and the way he was flying through the bush, got out of control and rolled it.

The next day three Buffels were sent out to do some or other recce. We were travelling through the bush and heard a chopper in the distance. They must have seen us and called us on the radio, saying they were attracting some small arms fire and directed us towards it. As we neared the area, we jumped out and split into two sections. I was leading the group on the south side.

The gunship was circling and occasionally firing into the bush ahead of us. When it did, you could see the chopper lurch from the 20mm cannon's recoil. I was moving ahead of my team and noticed movement on the other side of a small clearing. Diving for cover, I shouted something in Afrikaans to make sure it wasn't our other team whom we had split from as we disembarked. There was no answer so I let loose a few rounds into the area. I noticed a bit more movement, and then nothing. The gunship fired a few more rounds quite close by and there was some more movement further away, which was also rewarded with more rounds from my R4. Once again there was a bit of movement.

I was just about to get up and cross the small clearing when I noticed one of the section two troops trying to get my attention. They had now caught up, and he said there was a guy in the bush on the other side of the clearing whom I had not seen. I asked him where, and he pointed out the area. As I aimed my rifle at the area he asked if he could have this one. I said go ahead, which he did.

The chopper landed in a clearing near the Buffels and the five FAPLA bodies were brought together. We took a few rifle belts and magazines as souvenirs and the chopper guys took some AKs. The FAPLA soldiers were only youngsters but then, so were we. We were in the middle of nowhere and the chopper guys said they were not going to take the bodies. We said, '...*neither are we.*'

I don't know who decided, but we gathered all the wood we could find lying around the area, made a big fire and solved the problem. The one guy in section two who got our team's third kill was watching the fire and said: '...*Ons het hulle afgekoel, nou maak ons hulle warm.*'

John and his bicycle on midday break in Ovamboland.

26
Another One Bites The Dust

Mix youthful immortality, bikes and guns with a pinch of stupidity and what do you get?

Blood, lost skin and broken bones but most of all, lots of laughs. We all crashed, some harder and more often than others. My platoon in Ovamboland had two serious crashes that resulted in one broken leg and one failed decapitation. Other than that, they were the source of much amusement. We were merciless about stupid or show-off moves that resulted in some measure of pain and/or humiliation. Once in a while your own clever stunt didn't work out as planned or you ran out of talent at a critical moment, but such was the price of admission.

In May both platoons were driven to a transport depot in Grootfontein, where we collected 70 brand new XR 500s. They were the new ProLink single shock models, painted brown and mostly stock. Unlike the older 'bigwheel' XRs with the 23-inch front wheel and twin shocks that had huge bashplates, brush bars and light guards welded on, our new bikes only had two small modifications. Each bike had a luggage rack mounted on the rear frame behind the seat, and a round piece of flat metal welded to the foot of the side stand. An ingenious bit of applied physics that prevented the bike from falling over when the side stand dug into the sand.

We were very chuffed with our new rides. The older bikes didn't handle very well to begin with and the added weight really turned them into pigs. The bashplates were ankle-busters and if you ever went over the handlebars and hung on, thinking you could save it, the brush bars would break both your wrists. We didn't like them much, but didn't bitch too loudly. Riding a bigwheel XR was still a higher calling than trapping around the bush like some infantry twat.

The new bikes were really brand new. Mine had 0,5 km on the clock, the highest mileage on any of them was 4 km. Despite many warnings to take it easy and not fuck around, a white-haired troop named Seb moered down on the tar road turning left out of the depot. Total mileage before first crash, 0,3 km.

After the last *Volcano* terr was accidentally captured by a

Bike park at SWASPES, May 1982.

farmer and his son out hunting near one of our improvised motocross tracks, I was transferred from platoon 11 to platoon 12. With a new loot just arrived from Berede, we took over from the July intake leadership who were headed home. We spent some time doing

Training in the gravel quarry just east of SWASPES base.

training and maintenance with the new bikes and new platoon, which meant we basically rode all day every day. We found a gravel quarry next to SWASPES and a sandy MX track 40 kays away at Kombat mine that provided excellent training and many thrills and spills.

We soon discovered that 'bike maintenance' was a convenient excuse to gyppo out of the kak of army routine. At least for the troops. From my platoon sergeant perspective it was an easy, non-violent way to control 30 hooligans. I had an understanding with the platoon from

Sharpening my sand riding skills at Kombat MX track.

the start: as long as they didn't cause shit that got *me* in trouble, I'd play the maintenance card to get them out of as much digging,

Training ride along the Tsumeb railway line. My bike in the foreground was a 1981 model. On the right a 1982 model, the only difference was the shape of the front fender. Note the huge speedos, which didn't last long once we started wheelying.

sweeping, scrubbing and painting as possible. They tested the terms of agreement once or twice, but overall it was a classic example of mutual benefit. Later, at Ogongo, it worked just as well except that time the wily RSM had us figured out. To his great credit he played along and left us alone.

Minor tumbles aside, things went relatively smooth and bloodless. We did quite a bit of riding on railway service roads, especially between Otavi and Tsumeb. We came to know that one well. I had previously ridden it at night under more serious circumstances. The first time we rode it by day, two riders named Reeves and Welsh repeatedly played the old *falling-behind-and-racing-to-catch-up* game. I warned them several times that they would regret it before the day was out. They of course lagged me af, so I left them alone to learn their lessons.

The railway service roads were good tweespoors that followed the line and switched back and forth across the rails where terrain dictated. We came to one such spot where the track went between some bushes and made a sharp, blind left turn over the rails. We all immediately knew *this* was going to be a laugh. I hid behind a bush

Johnny and his pal Reeves overshoot a S-turn along the Tsumeb-Otavi railway line at 80 km/h. Note the audience at top left, and that neither is on the brakes yet.

with my camera while the rest of the platoon waited and watched on the other side of the rails.

Sure enough, our two racers came screaming between two bushes and never even saw the turn. I heard them swearing as they went by me doing at least 80. They donnered a long way into the bush and one ended up on his arse just as I had predicted. They were lucky to get away unscathed, and the rest of us were in stitches. I didn't bother saying '...*I told you so*'. The sheepish looks on their faces after they got themselves extracted from the thorn bushes and rocks and got back on the tweespoor was reward enough. I snapped a photo as they went past me, and months later when I had my films developed discovered what a classic moment of stupidity I had captured for posterity.

After we got used to the new bikes, wheelying was top of the list of higher skills to master. Inevitably most okes flipped their bikes. The XR 500 had a speedo the size of a soup can that stuck up above the handlebars and was the main point of contact when a wheelie went bad. It was also one of the most expensive parts on the bike.

The Captain in charge of Spes base was a funny oke in his mid-twenties who tried, and failed, to be a hard-arse. He was noisy, but

Wheelie heaven on the wide open shonas south of Oshakati. Part of 14 bikes line abreast on the rear wheel.

his bark was much worse than his bite because he was basically a decent type. He jumped up and down when he saw the first flattened speedo and yelled that we were fucking up his XRs, so from then on wheelies were strictly *verboten*. Of course, we lagged him af and kept wheelying, and fortuitously the platoon was deployed to Ovamboland soon afterwards.

The wide open shonas of Ovamboland was wheelie heaven. You could ride around on the rear wheel for as long as your skills and nerve allowed, with no worries about hitting anything. Except the ground, of course. With plenty of opportunity to practice, we all became very good at it. It was very satisfying to feel that magical balance point, shifting through the gears until you were doing 80 or 90 on the rear wheel. That bred an inflated sense of skill and confidence which eventually, naturally, reverted to humility for some.

Like me.

I came back to earth one Sunday morning on a huge shona south of Oshakati somewhere. We'd spent the previous afternoon in that area riding and wheelying in formation for the benefit of those with forbidden cameras. The loot was in the lead, cruising along a hard-as-concrete tweespoor on the edge of the shona. Most were on the rear wheel, including me at the back of the pack. I was having a ball, very impressed with my own skill and daring.

That suddenly ended. One second I was a moto god, the next I was sliding down that hard dirt track on my back with my shirt, webbing and R4 bunched up around my neck. I liked to leave my shirt hanging out for ventilation, but skidding along on my bare back wasn't part of the plan. I was flat on my back at 80-plus by the time the XR's speedo touched the ground. It was flattened to a third of its normal size. The glass shattered into long rectangular shards that made a tinkling sound as they came tumbling towards me in slow motion. Small red bits from the broken brake light added some colour to the scene. All this was accompanied by the loud scraping noise of the bike sliding down the track. Fortunately I couldn't hear my back being exfoliated, in fact I didn't even feel that till afterwards.

When I finally slid to a halt, I lay dazed for a second or two, then sat up. I remember saying '...*fuuuuuuck me*' and taking off my helmet. Next thing I knew, I woke up on my back, my helmet next to me and not a fucking soul in sight. Those bastards hadn't even seen me go down.

I slowly and painfully got back on the bike and rode off in search of friendly forces. I found the platoon five kays down the road, scattered in a circle around a big tree. It was tea time. They were lounging around in the shade, making coffee and talking kak. Dusty and bloody, I slowly rode into this circle with a flattened speedo and bent handlebars. I got the standard reception. They rolled around laughing, pointed at me, examined my back, and generally had a grand old time at my expense. Any sympathy I may have kept in reserve for future crashes by these jokers instantly evaporated.

A rough Sunday morning near Kwambi, south of Oshakati. An hour earlier I had flipped my bike at 80+ when a wheelie went bad.

After the hilarity died down, all I wanted to do was sleep. In hindsight I realise I probably had a concussion, but my bloody back hurt worse. It became the biggest fucking roastie I've ever had. The muscles along my spine were two red stripes and much of the rest of my back was raw. I lay down on the softest surface I could find, which happened to be my bike, and went to sleep. Some clown took a photo of me, with my own camera *nogal*, while I was snoring. When I look at it now, it makes me smile despite the momentary twitch of muscle memory in my back.

That flip baffled me for a long time. One minute I was the Doug Domokos of Ovamboland and the next, roadkill. Years later when I understood the dynamics of wheeling better, I concluded that the front wheel had stopped spinning from being off the ground for so long. The gyroscopic effect of the spinning wheel stabilises the bike and when it stopped, the balance point changed. Regardless of exactly which law of physics bit me in the arse that Sunday morning, the results were typical: one oke bloody and bent, the rest cracking themselves.

Angolan 'Daytona' where I could only take photos because my left wrist was badly sprained and I was riding one-handed.

The slight setback didn't stop me from wheeling. Or flipping. Some time later I did it deep inside Angola. I was going slow in second gear when I looped it, and stupidly stuck out my left hand to break the fall. I got up holding my wrist, and watched it swell to twice its normal size right in front of my eyes. In less than a minute it was as stiff as a claw. I could barely flex my fingers. Not to worry. I bandaged it tightly and practiced pulling in the clutch from the shoulder with my useless arm and four stiff fingers, and was ready for battle. I rode around Angola for several days with my left arm in a sling.

After a while it was easy, and it became one of the most thrilling rides I've ever had. The loot up front still rode as hard as he could. I kept up one-handed at the tail end, winding through the mopani trees at high speed, flicking the bike left and right and leaning into the small berms formed in that sandy soil by the 20-odd bikes ahead of me. Pulling away from a stop was a bit rough. I had to wind the throttle and drop the clutch by sliding my badly sprained hand off the end of the handlebar, but it worked well enough considering where we were. I did miss out on another thrill though. We came across a round earthen dam that had been breached on one side and formed a banked circle, the Angolan version of Daytona. The surface was rough and rutted and I couldn't keep my bike straight with one hand, so I took photos while the other okes went around faster and faster.

We stopped at Okalongo one fine day to refuel. The base was buzzing with activity. Some big operation was about to start and Buffels and Casspirs were lined up all over the place. More than two hundred troops and Koevoet were sitting on their vehicles, watching us as we left. Some okes were *windgat*, popping wheelies and raising the powder-white dust. All very heroic and impressive. We always drew attention, and we tried not to embarrass ourselves in front of strangers.

Some didn't try hard enough. Rifleman Steyn got the bright, make that *sheer fucking genius,* idea that wheeling out of the base from a standing start would really impress the big audience. In the excitement of the moment, he forgot that skill and bravery aren't necessarily the same thing. He dialed in what he thought was enough throttle and dropped the clutch. That XR snorted once and flipped right over onto its back on top of him. I swear, he bounced up and over onto the wheels so fast it looked like a video rewinding. Steyn was a shy blonde oke and not used to being the main attraction. I could see his red face glowing like a flare through the white dust as he jaaged out the gap in the sand wall with his tail between his legs. I was sitting there waiting for everyone to pull out ahead of me and laughed so hard, I fell off my own bike. A handful of his platoon mates also witnessed his cunning stunt and we teased him mercilessly until someone raised the stupidity level and took the spotlight off him.

Everybody got their share of ridicule. The loot was always out front, charging through the bush at high speed. He seldom crashed but when he did, he applied maximum enthusiasm. He came short in spectacular fashion one day after an argument with one of the big anthills that dotted Ovamboland.

Some of those things were huge, four or five metres high, and supposedly they all leaned north. They had conical bases with tall towers on top and apparently consisted of clayish soil, because the PBs made bricks with it. For some reason they quarried square holes into one side of the base only. From the other side the anthill looked perfectly normal. We soon learned that launching off the bases of these anthills could spice up your day a wee bit more than you had planned. We had a few minor scares that way but the anthills were like magnets to us, starved of elevation changes in that flat landscape. Therefore, the lessons were not learned and the loot delighted in leading the long line of bikes jumping over anthills. The closer to the tower you hit it, the higher the jump and the further you launched.

The platoon skidded to a halt ahead of me one day. By the time I got up front the okes were rolling around laughing. The loot was slowly getting untangled from his bike at the bottom of a two metre deep hole in the back side of a big anthill. He had hit the base of the anthill with impressive velocity and didn't even see the hole until he was in the air. We were in hysterics for a while, then a few okes climbed into the hole and lifted the XR out. The loot was red-faced and angry but was soon laughing along with us. Not much else he could do.

Racing around Angola one day, I saw the familiar dust cloud and bikes bombshelling left and right. One of the section leaders, Corporal Fourie, clipped a sawn-off tree stump with his footpeg and got pitched off the bike at 80 plus. It looked like he set off a silent

Section leader Fourie standing with arms folded, gathering his wits after taking a hard fall in Angola north of Dombondola.

landmine. His kit was spread over a 20 metre radius. Fourie was not a happy boy and when I laughed at him he got very unimpressed with me. It took him half an hour to gather his wits and scattered kit, and he wouldn't talk to me for a week afterwards. But hey – what was I supposed to do? It was fucking hilarious, and we all knew the rules.

Everyone eventually pays the piper. Twenty years later the big one caught up with me. Soon after my 40th birthday I went riding in Baja Mexico with three friends. It's a beautiful, rugged and wild part of the world, much like Africa in many ways. An hour from the finish, I purposely fell behind to stay out of the dust and was racing to catch up on a tweespoor in the middle of nowhere. I knew better than most how that usually ended, but it had been a long ride and we were tired and close to home. I ignored my instincts.

A blind, off-camber corner just over the crest of a hill jumped up and bit me in the arse. I crashed my XR 650 hard enough to put me flat on my back, in a wheelchair, on crutches, and off work for nine months.

Nobody laughed at me this time. I was a bit disappointed. They would not have made it in bike squad, sorry to say.

Riding in Baja Mexico 20 years later with my friend Jody from Cape Town at centre.

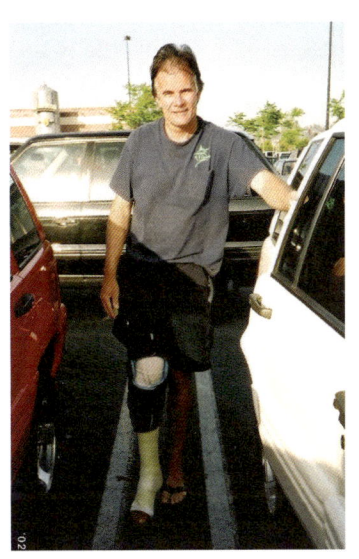

The result of playing falling-behind-and-catching-up games at high speed in unknown terrain in Baja. I thought of Reeves and Welsh while doing so, and crashed while I was contemplating slowing down. Off work for nine months, a life-changing event.

27

Six Foot Four And Full Of Dave

The platoon I went to Ovamboland with had some interesting characters. Dave was one of the shit-stirrers who always seemed to have a finger in the pie with any kak that went on. He was from Durban, one of the best riders in the platoon and an incurable wheelier. He had thinning reddish-brown hair and grew a long fu-man-chu moustache that hung below his brown pisspot helmet and dark Scott goggles.

Despite his rough exterior Dave was a good oke, dependable and fun-loving. He spoke a strange mix of English and Afrikaans slang that made sense to us, but must have sounded like gibberish to anyone outside our little circle. I can still hear him complaining about some perceived injustice I perpetrated against him once: '..*hey nooit kopperaal, blind sanctions man....why me*', on and on.

6 foot 4 Dave playing with a local boy's draadkar at a kraal north-east of Mahanene.

We interacted with the local people a lot and our relations with adults were usually friendly but formal. We realised that some of them probably had information about SWAPO's whereabouts, but we weren't trained in interrogation techniques and our random questions were mostly useless. Many, if not most, Ovambos spoke Afrikaans and it never ceased to amaze us that people deep inside Angola did too.

The kids were like kids anywhere, curious and playful, if shy. They were always attracted to the bikes and we gave them ratpack sweets whenever we had any. Young boys often wanted to go for rides, so we took them for a spin when circumstances allowed. Dave and his connection Colin from Margate went one better. They wheelied up and down with boys on the back of their bikes hanging on for dear life. It was a huge thrill for them and their watching mates, and we had a good laugh at their expressions and big white eyes. We even let some of the older boys shoot our R4s. Dave got to play with one lightie's *draadkar* in return.

We often TB'ed in the dry hollows that became small lakes in the rainy season, because they provided excellent cover. The bottoms of these were hardened dark grey mud covered in hoofprints and scattered dry leaves. I was up at first light one day, walking around the TB taking photos and enjoying the peace and quiet. I could hear murmurs of conversation from a few okes who were up and about, making coffee and packing their kit. Some donkeys stood under the trees around the hollow, quietly going about their donkey morning routine. We enjoyed them because their haw-hee-ing at night and in the quiet early mornings, and riding them, were sources of much amusement.

So, when a noisy three-octave ten-second fart, the kind that chases dogs off the porch and makes small children cry, ripped across the TB I assumed it was a donkey and had a good chuckle. Then loud laughter broke out at the far end of the TB 30 metres away. It wasn't a donkey, it was Dave crawling out of his sleeping bag and greeting the new day. The okes around him were cracking themselves and holding their noses. In the quarter-century since, I have *never* heard anything remotely like it. Never before either, come to think of it. Dave certainly had some unique talents.

His real genius lay in the field of music. The Aussie band Men At Work had a big hit at the time with their song *Land Down Under*. We played it endlessly and never got tired of it. One stanza went:

Sunrise in a TB east of Okalongo, with my recently vacated bed at far left.
Minutes later 6 foot 4 Dave crawled out of his fart sack at far right and
greeted the new dawn.

'...Buying bread from a man in Brussels,
he was six foot four and full of muscle....
I said do you speak-a my language,
he just smiled and gave me a vegemite sandwich...'

Dave was modest and a bit insecure, so he changed it to '... *he was six foot four and full of Dave'.* It became our anthem and every time it played, we belted out the Dave version with great enthusiasm.

Dave was the last member of the platoon I saw in uniform. I gave him a ride from Potch to the

6 foot 4 Dave helping prepare lunch in Angola, above beacons 9 and 10.

Joburg station the day the platoon klaared out and we all went our separate ways. He told me he was headed for Miami, where he had some family connection.

I often think about him and laugh. When I hear that song I see him as clearly as if he is standing next to me, grinning slyly about some gyppo he pulled and got away with.

28
Brommer Rides

Brommer was a surfer from Margate, on a different frequency than the rest of the okes. He got the nickname from his habit of making a strange *v-v-v-v-z-z-z-u-u-u* sound at the end of his sentences. Tall and gangly, he had spiky brown hair and pale blue eyes in a long angular face usually broken into a crooked smile. I never quite figured him out, but he was a nice and easygoing oke. Brommer and his Margate connection Colin, a one-liner, were usually together. Colin was one of the platoon's wheelie addicts and a very good rider.

A 101 Bn troop was assigned to us as interpreter for a series of one day patrols. He wore black takkies, a sheepskin-lined flying helmet thing that made him look like a cartoonish Red Baron, and carried a G3. He was a friendly oke and seemed chuffed about riding on the back of a motorcycle for a few days. Brommer was designated as his chauffeur for the first day.

Things went well all morning and we stopped at several Ovambo kraals. With the help of the interpreter we were able to communicate much better than usual with the inhabitants. Around mid-afternoon we rode a section of sandy tweespoor approaching a kraal. Besides the time we came so close to disaster at Ogandjera, it was the only instance of riding a tweespoor that I recall. We avoided them like the plague, for good reason.

This section was quite a thrill though. Where the tweespoor turned, high berms formed in the soft sand and we had a jol railing them, hard on the throttle. All except Brommer. He was having trouble with the extra weight on the back of his bike and bliksemed down twice in less than 100 metres. His passenger was understandably not enjoying his ride anymore. Neither was Brommer, so they parted ways by mutual consent. Colin, being a better rider, resolved the problem by taking the passenger on his bike. We were all pissing ourselves laughing.

Even the 101 oke was smiling but he actually wasn't all that amused, it turned out. He didn't pitch the next day and we never saw him again.

Brommer's passenger after he got onto Colin's bike.

29
Your Turn Is Your Turn

While operating out of Ogongo, I got orders from our old enemy the PF Intelligence captain one evening. I had to sweep the cutline between two beacons north of Okalongo with a section of bikes at dawn the next day.

I protested vigorously. We had demonstrated before to some idiot camper colonel at Tsintsabis that it was impractical. You just *can not* ride a bike slowly in thick sand. We wouldn't see a fucking highway across the cutline, never mind SWAPO spoor. What concerned me more, it's impossible to detect any sign of landmines in the deep sand.

All to no avail, the PF doos knew best. We went to sleep knowing something bad was going to happen. The area was rotten with swaps, and weekly landmine explosions kept everybody on their toes.

Some time during the night, a clerk woke me up and said the plans had changed, we didn't have to go anymore. I said OK, rolled over and went back to sleep. Around tea time the next morning a commotion broke out at the HQ. I walked over to see what was going on. A crowd was gathered around a Withings that had just pulled into base, towing a Buffel with its front end blown off.

Turns out it was the one sent to sweep the cutline instead of me and ten bikes. It had hit a cheese mine in the deep sand on the cutline. I just stood there staring at it in a blind rage. That fucking mine had my name on it but instead of bits of me decorating the trees, the Buffel driver got badly hurt. I felt a powerful urge to go beat that captain senseless or dead. After that my disdain for that stupid fucker turned to pure hatred that lingers, 25 years later.

Once in a while I still wonder what sequence of events occurred that night while I was sleeping that caused a Buffel's front wheel instead of my bike's to set off that mine. The slow, deliberate way it was played out like a chess game or some 1970s disaster movie still bothers me.

Who decided that it wasn't, after all, my turn? And why? Fuck knows.

View from the rear of a single file formation in Angola, November 1982.

30
Chicken Run

Besides Dave and Brommer, there were several other amusing characters in platoon 12. Truth be told, every one of them was a character in his own way. Some just seemed to be on centre stage more often when shit happened.

One such lightning rod was Johnny Welsh. He and his connection Reeves frequently attracted unwanted attention, sometimes through no fault of their own. Reeves was a tall, skinny rake of an oke with freckles and curly black hair. Johnny was shorter and muscular with blue eyes and a wry smile. Oddly enough they were both dead shots with the *snotneus*, so they each carried one in addition to their R4s.

Like fish and chips or Red Heart and Coke, Reeves and Welsh were never far apart. Like the day we rode the railway service road between Tsumeb and Otavi, and they lagged me af when I told them repeatedly that falling behind and racing to catch up would lead to tears. Luckily I was there with my camera when they overshot a corner and fucked into the bush as I had predicted. The look of annoyed surprise on Reeves's face as he got back on the road with the whole platoon laughing at them, was priceless.

Despite all our previous experience, we learned some new lessons on our first few patrols in Ovamboland. We wore bush hats instead of helmets a time or two, which made Johnny our first casualty of note. We were cruising along the side of a road doing about 70 when I saw a big dust cloud ahead and bikes bouncing around like ping-pong balls. Before I could react, I hit what I can only describe as a *gnarly whoop section,* or *chicken run* as we called it in those days. It was a dormant *mahangu* field ploughed at a right angle to the road. The furrows were a good half-metre high and rock hard. How I stayed on the bike, I have no idea. Even more amazing, so did everyone else.

While we stood around laughing and shaking after the big skrik, Johnny suddenly became the centre of attention. He was bleeding like a stuck pig, with bright red blood streaming down his face and neck. He scared us for a moment, but we soon realised that he hadn't been shot.

When he hit the Ovamboland chicken run, Johnny's R4 and the *snotneus* he carried flapped up and down on their slings and donnered him repeatedly on both sides of the head. He had several deep cuts to the scalp which caused some spectacular bleeding.

Two okes cleaned Johnny up and bandaged his head. He looked like an Indian snake charmer with a huge white turban. We all laughed at him, but the lessons were noted. Next time out we all wore helmets, and we kept a sharp lookout for old *mahangu* fields after that day.

31
Flor De Luna

Music was a big part of our lives in the army, especially in the bush. Certain songs instantly take me back to a different time and place when I hear them.

Back at Berede in Potch, Queen's *Another One Bites the Dust* was considered the official bike squad song. The clerks always had a radio playing in the bike office and every time it came on, they or Captain Stroebel would turn it up loud. I remember walking into the mess at lunchtime, bloody and dusty from crashing bikes, and having the berede loots and corporals sing it while pointing and laughing. A hardy bunch they were, kept their sense of humour despite smelling like horse piss most of the time.

In Ovamboland three songs were forever imprinted on my brain. *Land Down Under* by Men at Work takes me back to the shonas and mopani trees, and a hilarious character named Dave. That's a story all on its own.

I was sitting in the pub at Ogongo, brain in neutral, drinking a cream soda one afternoon. I'd gyppo'd some nonsense our old enemy the Intelligence Captain had going on. PT, if memory serves. A tape was playing behind the bar. The line *'...and now you find yourself in eighty-two'* made me sit up and take notice. I thought, *'...what a coincidence, a song about the year we're in'*. I vividly remember the barman, he was a big friendly oke with longish black hair who wore black PT shorts and a white T-shirt.

What burned that line into my memory was, minutes later that PF doos found out where I was and gave me and the loot a sandbag opfok in the hot sun and bottomless sand. Turns out the loot had been gyppoing too. The name of that song was *Heat of the Moment*.

One nice evening we were lying around talking kak in our tents at Ogongo. The tent flaps were down, the moon was bright and it was very peaceful. Music started drifting in from a nearby tent, a beautiful guitar instrumental piece. It was a haunting tune that made my neck hair stand up. I made an oke named Luwes volunteer to go find out what it was. He came back and said it was *Moonflower* by Santana.

Afterwards, I looked for that song for many years and only found it recently. It usually goes by its Spanish name, *Flor de Luna*. Twenty-five years later it still gives me goosebumps. I instantly see the moonlight shining into that tent, the shadowy shapes of the okes around me and Luwes' face looking in at us. When I think about it really hard, I am almost certain it was the night someone up the chain of command made a decision that caused a Buffel to hit a cheesemine the next morning instead of me on my bike.

In 2000 I was an airline pilot in America, and flew Carlos Santana and his wife from Las Vegas to Orlando twice. Each time the flight attendants told me what a nice oke he was, humble and polite despite his fame and genius. I wish I had shaken his hand and told him about the impact his song has on me. I think he would have appreciated it.

Plotting our position during a midday siesta somewhere in Ovamboland.

32
Danger White Phos

Before patrols or other operations, one of my main duties as platoon sergeant was drawing ammo. We carried lots of it: seven or more magazines of 5,56 mm ball and tracer for our R4s, 40 mm rounds for the *Snotneus*, M26 grenades, various coloured smoke and white phosphorous grenades, and sometimes claymores. We liked the 50-round R4 magazines loaded with 45. Any more, and the spring compressed too much and the R4 would jam. The first time that happened, it shocked the shit out of us because the thing was supposed to be unjammable.

White phos grenades did not go back to the ammo store after patrols. We had standing orders to explode them a safe distance away when returning to base. They were extremely nasty devices, capable of inflicting horrific injuries. You had to be very careful with them. They looked like smoke grenades except for the white band around the canister and the letters WP.

Back at Berede in Potch was a horse Corporal who had been burned by a white phos grenade. A popular stunt at the time was taking the parachute out of an illumination mortar round and replacing it with an armed WP grenade. At the top of its arc the grenade would pop out and explode in a thousand white streamers. I forget what went wrong, but the poor oke basically lost his hands. He couldn't bend what was left of his fingers and his upper body was badly burned.

Returning to Ogongo from one patrol, I collected the half-dozen WP grenades the platoon carried and chucked them into a ploughed *mahangu* field next to the road. The rest rode on into camp, I was all by myself throwing grenades and having a jol. You had to put some muscle into throwing the bastards because they had an unpleasantly wide lethal range. It was morbidly fascinating watching the white explosions with little bits of burning phosphorous flying all over the place.

I was back on my bike when I noticed that the grenades had set fire to the dry leaves and stalks left over in the plowed field. Not only that, but it was spreading towards the grass and trees on the far side of the field. I jumped over the fence, ran to the fire and started kicking sand onto the flames.

It took quite a while. I was trying to extinguish not only the dried remains of the last crop, but lots of burning phosphorous too. At one point I was losing the battle and considered riding into Ogongo to get some help, but realised it would be an uncontrollable blaze by the time we got back. I was very busy for twenty minutes or so. I knew I'd be in deep shit if I burned down half of Ovamboland.

The flames eventually died after I had practically ploughed the fucking field for the next season's crop. There was as much dust as smoke in the air by the time I rode into camp, pouring with sweat and even dirtier than usual. When I finally caught up with the platoon, they just laughed at me.

Lessons learned.

In Angola above beacons 12 and 13. L to R: Welsh, Le Roux, Steinvaart, Reeves, Heath.

33

The Great Cuca Robbery

Cuca shops were officially off limits to troops on operations. Once we discovered that you could get an icy cold Coke or Fanta from a small corrugated building in the middle of nowhere in the 40-degree heat, that was lagged af too. Just like *no cameras allowed* and *wheelies forbidden*.

Cucas varied from small corrugated shacks to large brick buildings. The variety of merchandise they sold always amazed us. I don't recall buying anything from cucas but cold drinks or the occasional tin of spaghetti meat balls.

Returning from Angola one day, we came across a suspicious sight: a deserted cuca with the roof caved in and one wall half destroyed. We approached it very carefully. After finding no spoor of any kind the loot, myself, and a few troops searched the inside. Something was not lekker. Several shelves were still stocked as if housewives were standing in a queue outside, others were knocked down and obviously looted. We found everything from tinned food to laundry soap to children's clothing. The loot and I smelled a rat and ordered the troops not to touch anything. We reported our find by radio to HQ at Ogongo, which later turned out to be a good move. We were baffled, but continued on our merry way and soon forgot about it.

The next morning around tea time we rode into Ogongo base. In passing I noticed an old Ovambo man sitting in the shade on a bench outside the admin office, but didn't think much of it. I saw his left profile. He wore khaki trousers, a brown blazer and a hat, and his hands were resting on a walking stick planted between his feet. He was obviously a person of some importance, but what struck me was the hostile expression on his face.

After lunch the loot was called to the CO's office. There was a big stink going on. The sour-faced old man had brought a charge of theft against the 'motorcycle soldiers'. The CO, according to the loot, was slightly amused because we were not the first bunch he'd accused. The story we got was, his cuca had been raided by SWAPO months before. And ever since, every time an army unit went by the cuca he would rock up at Ogongo, throw a tantrum, and demand damages. The army humoured him to a certain extent, paid him some money, then investigated the charge. He was obviously a sly old cunt who had the system figured out. That's what made me the moer in. They knew he was a crook, but still treated us as if guilty of his allegations.

The next day some MPs, two or three PFs and the whole platoon drove out to the ruined cuca in four Buffels. The old bastard got a free ride thrown into the deal. The MPs sniffed around and acted important for a while, asked a few questions, and we all drove back to base. A whole day of rondfok. By the time we got back to Ogongo late that afternoon we were not a happy lot. Those idiots had wasted a full day of our time, a day we could have spent riding and chasing swaps.

We were cleared of the charges, of course. But the platoon was violently angry about the affair because of something one of the MPs told them that evening.

That old fuck had accused us of stealing a pencil and one pair of baby panties.

34

Front Toward Enemy

A berede platoon shared Ogongo base with us at one stage. One day corporal Horn, the chubby PF platoon sergeant whom I knew from Berede in Potch, walked into the mess with little black dots all over his face and neck. He looked like he'd been spray-painted through a mosquito net. Turns out he'd lost an argument with a claymore on patrol two days before.

Horn had set the claymore at the base of a mopani tree trunk and kneeled behind the tree while he connected the metre-long det wires to the wires leading back to the TB. But someone fucked up. The claymore went off the instant he touched the second wire to the det wire. He got a snoutful of Ovambo sand that blasted a thousand little dents in his face.

The loot and I laughed and laughed at him, but he wasn't amused. *Au contraire.* He got *very* annoyed with us. Horn was a gentle, good-natured soul and this uncharacteristic hostility baffled us. Only thing I can think is, a claymore backblast viewed up close causes a loss of sense of humour.

35
Heads Or Tails

One afternoon we ended up at a Koevoet base in Sam Nujoma's home village of Ogandjera, 40 kays south-west of Ogongo. The small base had been converted from a police station by bulldozing sand berms around the buildings and adding sandbag bunkers and barbwire. Just outside the base the sandy tweespoor track made a S-turn where fences channelled traffic through a narrow bottleneck.

The loot at Ogandjera after fences forced us to ride a tweespoor into the base.
He was not happy.

We stopped when we came up to it. Our formation had gone from the usual T to line astern as the fences closed in. We did *not* like what we saw. We tried to find a different way in, but the place was surrounded by fences and kraals.

Much of our daily routine was aimed at avoiding landmines. We never rode the same route twice. We avoided roads, especially sandy tweespoors, like the plague and kept an eye on the ground everywhere but the wide-open spaces. After some discussion, we decided the odds were 50-50. The loot flipped a coin and led the platoon into the base, riding on the sandy *middlemannetjie.*

It was a nerve-wracking 60-second ride. I remember that section of tweespoor as clearly as if I rode it yesterday. I have a photo of the loot taken just after we arrived. He is sitting on the sand berm smoking, seemingly relaxed, but his body language is not his normal easy going self. He looks tense and distant.

We spent the night, which I have no memory of. Early the next morning we were ordered to some other destination and had to leave before the sappers' morning road sweep. Same thing, 50-50 odds, this time we rode through the bottleneck in the tracks of the tweespoor. No decision needed, we'd used up the other option riding in the previous day.

A few days later in another base, someone walked up to the loot and told him that the sappers had found a cheese mine in the *middlemannetjie* that morning after we left Ogandjera. He said that a swap sitting on top of a nearby cuca shop had watched us ride in the previous afternoon.

36
Who Wants To Live Forever

Soon after bluffing my way into bike squad, I started wondering if it was such a clever move after all. Between December 1980 and March 1981 two loots and a section leader were killed. I suddenly realised that I might not survive three years in the unit, but I was determined to have a good time either way. At least I wouldn't have to *trap* around the bush like a doos.

On 5 January 1981 bike squad suffered its only combat death when section leader Pieter Swanepoel of platoon 6 detonated a cheese mine on his XR 500. I arrived at Berede on that day and heard about it shortly afterwards. While drawing my rifle and webbing at the weapons store the chatty clerk showed me Swanepoel's R4, which had been returned to Berede.

Or rather, what was left of it. The folding stock, pistol grip, trigger and hand grips were gone and the barrel was bent. The steel body was scorched matte black and looked like it had been burned in a fire. It made a deep impression on me. I stood in the dimly-lit store holding that R4, trying to imagine what the force that bent the barrel like that did to his body. I found out two years later, when I became close friends with the oke who was riding behind him when he hit the mine.

A week or two

The remains of Swanepoel's XR 500.
Note bent R4 at left.

before I arrived at Berede, a loot was killed when he hit a garbage truck on a Honda CR 250. His name was Burger, I vaguely recall that he was the platoon 10 pelbev. He raced the motocross nationals and had his own bike in camp. A dirt road led from the back of Berede to the training areas and MX tracks. About two kays out it made a left dogleg and ran between two rows of trees. The garbage truck, a tan Bedford of 1950s vintage, was chugging along at the end of the lane right as Lt. Burger came flying around the corner at high speed. He hit the truck so hard that his upper body dented the curved bonnet in a good 30 cm. Those old trucks were made of fucking steel plate. He didn't stand a chance. It stood in the vehicle park for a while and I saw it around the same time as the mangled R4. Another grim impression.

One of the new loots who arrived from Oudtshoorn with me was a nice oke from Cape Town named Neil Johnson. He was one of the Cape's top 125cc motocross riders and was looking forward to racing the very competitive Transvaal circuit. A few weeks later, during a race at Brickor in Edenvale, Neil crashed after the start and several bikes ran over him. He died in hospital four days later. I can still see him sitting on a table in the bike squad office, smiling as we talked. I also remember Kevin, my loot friend from Oudtshoorn whom I ran into at SWASPES in 1982, breaking the news to us. We didn't know it, but Kevin was living on borrowed time too.

My platoon in the bush was lucky. We all returned home at the end of 1982. We only suffered two serious casualties that had to be casevaced.

On one of our early patrols, we were cruising along at dusk one day. We were riding in T formation, with the loot towards the left flank of the line abreast, and me leading the single file behind. We crossed a shona and entered an area of scattered trees, many of which had been cut off one or two metres above the ground. We were doing about 60, looking for a good spot to TB. My head was on a swivel and looking to my left, I suddenly saw the loot skidding to a stop in a cloud of dust. The okes either side of him were doing the same so I instinctively jumped on the brakes with maximum enthusiasm.

A sturdy steel wire fence brought me to a halt. The second wire from the top was against my front forks, between the wheel and the fender. I looked around to see if everybody was OK. The right flank of the loot's section was abeam me. All except Luwes managed to stop. I saw it happen in slow motion. Luwes was looking straight ahead and never saw the fence or the other bikes skidding to a halt. He went through the fence doing 60. His bike broke the bottom wires and continued on, but the top one caught him by the neck. He did a vicious somersault off the back of the bike and landed flat on his back.

Everybody rushed over to help. Luwes was well-liked, always smiling and cracking jokes, and we were very concerned. He stood up slowly and painfully and started gathering his wits. He said his neck was sore, but after a few minutes he seemed to be his normal self and we all heaved sighs of relief.

Our first serious casualty rode through a steel wire fence east of Okalongo.
He is standing fourth from right, by the bike's front wheel.

While talking to Luwes we noticed that his neck was getting thicker. It puffed up like a balloon and something was obviously seriously wrong. The loot immediately called in a casevac. A Unimog ambulance soon arrived and took him and his bike to the nearest base, from where a chopper flew him to Oshakati. It took him a long time to heal. The next time we saw him was when we left Ovamboland in December.

Turns out the wire had torn his windpipe, and air escaping under the skin made his neck swell up like a frog's. It was a hard lesson about local customs and being wide awake at all times. The PBs cut down trees for various reasons and often used the trunks as pre-installed fence posts. After that incident we were very vigilant around trees, and never again cruised around fat, dumb and happy after sunset.

Our second casualty was another good oke, quiet and dependable and, like Luwes before him, sorely missed after he was casevaced. It was all a bunch of loudmouth 52 Bn Buffel drivers' fault. They bragged to us that they held the record between Ogongo and Okalongo, about 40 kays away. They had done it in an hour and twenty minutes and thought they were shit hot. It was an easy trip. You just went east along the tar road for a bit, jaaged north on the Okalongo shona and when you crossed the white road parallel to the cutline, there it was.

The gauntlet had been thrown. Leaving Ogongo one day, we decided to shut the drivers up once and for all. We fell into single file and the loot took the lead. As usual we rode alongside the main road, which slowed us down a bit, but once on the shona we *steeked* it. We knew the route fairly well, which helped. I saw 120 on my speedo most of the way. Twenty-six minutes later we pulled into Okalongo base, yelling, laughing and swearing from the very enjoyable adrenaline rush.

We were very chuffed with ourselves until someone said, '... where's Geeringh?'

Panic stations. The loot yelled at an HQ clerk to send an ambulance back down the shona and we tore out of the base, very worried about our missing mate. Halfway down the shona a PB flagged us down. Geeringh was lying under a tree with his head resting on his kit, his bent bike nearby. He had crashed and broken his leg and some friendly PBs had carried him into the shade. We were very relieved to find him in reasonably good shape. The ambulance showed up soon afterwards and he was casevaced to Oshakati.

Geeringh was the *moer in* with us. He had been towards the front of the weaving line and nobody saw him go down. Two thirds of the platoon, including me, had raced by maybe ten metres away without seeing the dust cloud or his frantic waving. He shot off a full R4 magazine with tracers every five rounds, but we were so intensely focused on the terrain at 120 that nobody saw or heard a fucking thing.

Years later it struck me that we must have missed many swaps the same way. I sometimes wonder how many ambushes we didn't trip and how many kills we didn't get, jaaging around like we did.

Hard to say.

Our second serious casualty, who crashed at high speed on the Okalongo shona and broke his leg.

37
Skulls

The only combat casualty bike squad ever had was Lance-Corporal Pieter Swanepoel, who hit a cheese mine near Oshigambo on 5 January 1981. I joined the unit on that date, and I was shown the remains of his R4 at the store when I drew my kit a few days later. Only the action and bent barrel was left. It made a big impression not only on me, but the whole unit and its way of thinking. We were more concerned about mines than any ambush, and took extraordinary measures to avoid more mine casualties.

Both bike platoons in 1982 had numerous close calls but no casualties, despite the best efforts of SWAPO and the 2IC at Ogongo. That dumb bastard seemed intent on killing some of us with sheer stupidity.

As platoon sergeant I didn't normally attend order groups before operations. My job was taking care of the logistical needs of the platoon like food, ammo, radios and tools while the loot went to order group. I did attend a few though, usually when there was big kak. I forget why, but I was present at order group for an internal patrol in the area south-west of Ogongo when they made a very big deal about dead cows. The *spies* had information that SWAPO was marking the locations of landmines by placing cattle skulls in trees. No mention of where the mine was in relation to the skull, or how the skull was placed, just the fact that a skull in a tree meant... *LANDMINE!*

Two days later we came charging off a shona into the dense mopani bush north of Ogandjera somewhere. Suddenly everybody locked up brakes and skidded to a stop in a cloud of dust. I always rode sweep at the back of the formation so I went up front to find out what was going on. The loot and the troops around him were sitting on their bikes, side stands out and feet on the pegs. I thought, '...*what the fuck are they doing*?' and stopped next to the leading section. The loot pointed at a tree close by.

In a fork two metres off the ground, a snowy white cattle skull with wide horns glared maliciously at us.

I about shat myself and did the same as everyone else. I put my side stand out, killed the engine, and made very sure my feet didn't touch the ground. By this time we were experienced bush riders, and having closed up the formation entering the trees, most of the platoon was within a 30 metre radius.

There we sat, too shit scared to get off the bikes.

The loot always kept the oke who carried the B25 radio close by, so he called back to Ogongo. He had several tense conversations explaining the situation to different people. The last listener was our old enemy the Intelligence captain, no doubt sporting his fucking shorts and Grasshoppers.

Meanwhile the rest of us were scanning the ground for any sign of disturbance. It was a waste of time. The soil in that area was the typical light grey, hard packed Ovambo sand and you'd never see any sign of a hole dug and filled in before or during the last rainy season.

Things got a bit tense to say the least, not helped at all by the verdict from the captain safely in the radio tent at Ogongo. He said it was OK, don't worry about it, keep going. The loot was a wild character, fun-loving, but short-tempered and very aggressive. He was arguing with this cunt back in camp, why had they made such a big deal about skulls in trees, and now that we were right on top of one it's nothing to worry about?

The argument went on for a while and he finally threw the handset down in disgust. No sappers were coming out to sweep the area. We were ordered to keep going. That caused a near riot. Some okes threatened to shoot that stupid bastard captain if one of us got blown up. I made it clear to them that they could only do that after I'd slit his fucking throat. Provided it wasn't me that got blown up, of course.

By this time 45 minutes had gone by and sunset wasn't far away. We were caught between a rock and a hard place. The loot and I looked at each other and shrugged. He fired up and led off past the skull in the tree. We all knew someone was going to die. I mechanically started my bike, clicked it into gear and waited for my turn to go, all without putting a foot down. I had a strange detached feeling, almost floating, and everything slowed down like single frames on a screen. We were all yelling and swearing, okes were wheelying, spinning wheels, and screaming abuse at the army in general and that captain in particular. I watched individual grains of grey sand launch off the rear tyres of the bikes pulling away around me, arc upward in extreme slow motion and disappear behind me. I was hard on the throttle, yelling something about *fuck the army* as I clicked through the gears. Through all the noise I could hear every metallic *click* as I upshifted.

We jaaged away from there like the start of a motocross race and didn't relax till we were a few kays away. We stopped again, got our pulse rates under control, and did a head count. After blowing off steam we rode on, a very pissed-off bunch. We never forgot about it and that captain became our mortal enemy that day. I still get fucking angry when I think about him.

He even made us run 2,4s in the sand at Ogongo. What a doos.

Platoon 12 in thick mopani bush in Angola, November 1982.

38
Untouchables

On ops in Ovamboland, my platoon got thrown out of every base we operated from except Ogongo. It amuses me now, but at the time it created a lot of resentment towards the company and battalion command structures we dealt with. The PF mentality was prejudiced against us simply because we rode motorcycles. Most of them saw us as *sleg moer* zol-smoking biker scum and treated us accordingly. The result was, we slept in the bush almost every night we spent in the operational area.

To be fair, some of it was self-induced, but what came first? The chicken or the egg?

We lasted less than five minutes at Okalongo. Riding into the base, we clearly saw several bags of vegetables and fruit hanging in the kitchen tent, caught in a beam of light like you see in romantic movies or shining unto Moses in an illustrated childrens' Bible. The loot and I were still checking in with the base commander, a SAI Major, when a commotion broke out behind us. Most of the platoon hadn't even dismounted before two of our okes got caught stealing oranges. We were promptly told to fuck off and not come back. I was annoyed with the two troops for a few minutes, but soon remembered where my loyalties lay so we just laughed about it.

The first time we arrived at Ombalantu, a movie was being shown inside one of the buildings. We pulled in, dusty and tired, wearing webbings and R4s. I was wearing a cut off T-shirt, sitting amongst my platoon on the floor watching the screen. Some SAI Major, presumably the base commander, walked in the door to my right and asked where the bike platoon sergeant was. When I stood up and said '...*here*' he started kakking me out and proceeded to get himself all worked up. I couldn't get a word in edgewise, but I could tell the man was not happy.

End of the story was, that cunt made us sleep against the outside of the sand wall around the base for the two nights we spent there.

We all knew if SWAPO revved the place with mortars we'd be very fucked. Lucky for us, and that Major, it didn't happen.

A piss lily with lime around its base at Ombalantu, right rear. In foreground, me and my high school friend 'Spotweld' whose cousin was our tracker mate Dave's girlfriend. Ombalantu was a bleak place, we didn't mind all that much getting banned from the base.

Getting thrown out of Mahanene took a bit longer.

A pisscat 3 SAI CSM named Sparks, a PF temporary Staff-sergeant, and one of the *dominees* eventually did it. Sparks took the ingrained PF dislike of bike squad to a new, alcohol-fuelled level and hated us for being alive. The *dominee* went on a moral crusade against us because the beautiful pub in the base looked a bit shabby one morning after the greatest piss-up in SADF history. The loot and I were only two in a crowd on the night in question, but bike squad was once again a convenient scapegoat. The CO of the 3 SAI company occupying the base was a former Angolan named Captain Rocha, a true gentleman, but he was shouted down by the other two and we were chased out within a day or two.

Perhaps for the best. Mahanene was the only base besides Ogongo in which we spent more than one night. We were pissed as fiddler's bitches every night of the cumulative week or so we spent there. It was hard on the body, even at that age.

Platoon 12 on the tar road at Mahanene.

39
Swannie's Story

After returning from Ovamboland in December 1982, I became friends with Patrick, a new loot who had been in platoon 6 in 1980-81 and went through Infantry School while I was in the bush. We are still close friends today. Patrick was riding behind Lance-Corporal Swanepoel when he detonated a cheese mine on his bike on 5 January 1981. This is his account of that day, in his own words:

A motorcycle can never set off an anti-tank mine, no ways, fucking impossible, we move too fast, weight distribution as opposed to speed. Ah fuck it!! I remember this dumbfuck officer briefing us the night before. After the order group he asks us, '...*what do you guys do if you hit a landmine?*'

HELLOOOOO POES PROPPIE, YOU DIE!! As a rifleman I restricted my comment to '...*one down and pop a wheelie*'.

In order to do justice to the memory of Lance-Corporal Pieter Swanepoel, I need to clarify one or two issues.

First, I write this account some 26 years after the event. Some aspects are as if they happened yesterday, and others are a confusion of time and distortion of facts brought about by the aforementioned. If I neglect to mention certain key people that were present it's simply because I don't remember all the details, forgive me and please send in the relevant corrections eg. what platoon were we in?

Secondly, I'm writing this story as it happened through the eyes of a national service rifleman as I was at the time. Subsequent to these events I joined the Permanent Force and in time was promoted into the officer ranks. To most readers this will mean that my views and opinions have become distorted over the years. NOT SO.

Lastly, no mention whatsoever can be found anywhere with regards the wonderful unit that we served in, as we used to refer to ourselves, 'BIKE SQUAD'. It's thanks to my good friend Platoon Sgt Yuri Maree that maybe the history of the wing might come to light.

If not for him and his consistent nagging, I would not be telling this story now.

We had been operational for upwards of a year, based in old Oshigambo, 53 Bn, deployed as a force multiplier and reaction force. The platoon had long before been split down to section level and sent off into different areas of operations. Constantly called out to do follow-ups, only to get very close and have to hand over to the paras to finish the job and get the credit for the kills. No offense to the paras, but guys, you have no idea how we hated you then, healthy competition. Our first loot, Lt Jansen had been pretty badly fucked up in our first contact and was not to be seen again. On the day of that particular ambush I was on orders in Ondangwa, having been caught some weeks before stealing sausage rolls out of the troops' canteen for our section. On entering the radio room I remember the sitrep coming in, Lt Jansen and rifleman Stephen Short had been wounded, NO KILLS. You have no idea of the ribbing I took for HIDING in Ondangs while the MANNE where fighting. Anyways, the Lt was replaced with a short service Sergeant, Clem Da Mata, really cool guy, good leader. However his time was short, and pretty soon he klaared out.

My bru Patrick Devy, who was behind L/cpl Swanepoel when he hit a Russian anti-tank mine on 5 January 1981. This photo was taken prior to that day, as evidenced by the presence of Sergeant Clem Da Mata in the background.

Sgt Clem was replaced by another 2/Lt (forget name) another cool dude, he was seriously fucked up, the man's uitklaar was overdue and he still stayed on. Eventually some twat figured this out and we lost him too. That left us under the command of section leader Lance-corporal Swanepoel.

Even though bike squad had a pretty dubious reputation, ie. Hells Angels, Durban drug addicts etc, we were serious about trying to kill the gooks. It was a matter of pride, plus we could go out on XR fucking 500s, get away from the jamstealers and do what we were sent there to do. What a life. Swannie attended the order groups, took the shit and away we went. And boy did Swannie take shit. Some fuckhead arrived in the camp one day as we were leaving for a follow up, this cunt couldn't accept the fact that we had R4s and strange webbing, this was most fucking unorthodox. Swannie had to explain to this prick that we were SWASPES and that this equipment was standard issue. This while the gooks have been located and we were the only guys in the area at the time that could catch them. (The PRICK was a BRIGADIER).

Due to the fact that we spent all our time in the 53 area and were based full-time at Oshigambo, we got to know the area very well. Needless to say all the 'usable' cuca shops were within our range, hahaha. Swannie had more shit to deal with after very hard follow-ups, we'd return via a cuca for a few snorts, haha. I can still hear Swannie, '...Kom manne, asseblief nee, kom hulle gaan my strip. Nee fok, ons ry nou'.

Hey man, Bikes were hard. Imagine nine, ten hours a day in Ovambo heat, webbing first line ammo, each carrying a 60 mm bomb, water and a 150kg bike between your legs, first and second gear all fucking day in beach sand!! Sometimes lifting a tracker. Needless to say we got pretty proficient at tracking ourselves. Of course we would partake in the odd drink now and again, not that Swannie didn't, but those are other stories. The fact of the matter is, we were disciplined.

It was around December 1980 that our section got shuffled for the last time. Some got lucky, some organised, who knows, it's just the way it worked out. I do remember sitting in the mess when we heard a fuck of an explosion. MINE, nothing new to us as we'd reacted plenty to this type of thing, calling in choppers for casevacs etc. Swannie got called in by the Company OC and off we went, first on the scene. What a mess, the funny thing was that as we arrived, won't bore you with the drills, we switched off the bikes and there was absolute silence. Nothing, not a bird, nothing. A Ford F100 was off the road on its side in flames, then we heard the crackling of the fire. It's strange how we went about the job, assessing the victims, all dead. I think there were seven. None of us spoke very much. Suddenly a baby cried. What the fuck, check the dead again, NO BABY. This apparition appears about 30m down the road out of the bush. It's hard to describe, but he's covered in dust and carrying a small baby, about nine months old, the child's making soft crying sounds. The only survivors. His brother was the driver, most of the dead his family. The child, his niece, they were on the back when they hit the mine, him no injuries, the kid a small wound in the chest. I think it was Swannie that brought in the Puma.

December in Oshigambo. Watching the SAI units come and go, making friends with insane fuckers looking for booze. (Not hard, we're bike squad). Fixing other peoples' fuckups, eg. jaaging down the Delta pipeline so we can tell 44 Para that yes, you have hit a mine on the pipeline, yes you are nowhere close to Oom Willie se pad. Jerry du Preez fucked his knee up on that one. The specific times and dates melt into a constant, yes, we were there and these things did happen. Follow-up after follow-up, 53 Bn area was very active at the time. When we weren't on follow-up we were on routine area domination or show of force patrols. Arseholes sending us out to deliver toilet paper to even bigger arseholes, how does a Lance-corporal tell a Captain that we are not there to supplement his inefficient logline, but would rather go to areas that we know the gooks move through? After all, we'd like to get some kills.

SCRAMBLE!!! SAKK (SACC) have been ambushed north-west of Echo tower. It would appear as if SWAPO gave them a hiding. Swannie runs out of the ops room (WE ARE READY), gives us the basics, and says follow me. We take the short cut through the mission station to join up with the old bus route that heads north to the Delta pipeline.

Now let me explain. When bike squad moves from base to a

contact scene, we move fast, the guys can ride, that's why we we're there, reaction force. When we get to the location, in specific terrain, we can track at approximately 15 km per hour. In this particular area, most of us could track without a qualified tracker. If we had a tracker, what a bonus. On this occasion we did, Sergeant *Rooibaard*. At that stage, trackers where in high demand and were practically run off their feet. These guys were so busy it was almost impossible to get one detached to you. We were lucky, we had one of the best. To pillion a guy at high speed through soft sand and thick bush is, well, say no more. To sit pillion and trust your life to some or other lunatic with an attitude plus an XR 500 takes big balls.

Swannie takes off a side track through some thick bush and suddenly hits the brakes! When moving like we used to, you automatically followed the same lines as the rider in front, when he stops, we all stop, or fall! Same thing.

We dust ourselves off and start bitching, '*...what the fuck, you wanna kill US or what?*' Swannie points at the track, it's got a small pile of twigs in the middle of one track, he says,'*...ek dink dis 'n myn*'. Rooibaard walks up and agrees, we radio it in and fly off, hook up with SACC and start the follow-up, sooo close but yet so far. Those gooks were pulling anti-tracking like us plebs had never seen. Rooibaard says no way, we've lost them. Heading back we meet with the sappers at the spot and they confirm they lifted a TM58, whew, well done Swannie.

That night we all get pissed and start reflecting on the mine issue. Often the sappers asked '*...how the fuck?*' we just laughed and said '*...what the fuck*'. We debate the issue seriously for the first time that I can remember as a section, and either that night or shortly afterwards Swannie said to us that he was going to die by a mine. He was morbid about it and asked his friends in the section to make sure that certain items went home. Fuck it, not going to happen!

We the willing
Led by the unknown
We who have done so much with so little for so long
Are now prepared to do anything for nothing.

It was a little rebellious ditty that we would recite from time to time, but only if we thought the PFs had screwed us.

Due to the fact that we had spent so much time together in the operational zone, things sort of fell into a natural order. The first rider out the gate normally led the section to where the tracking started. If you were in front, you didn't eat dust. Also, some of us knew certain routes better than others. If so, you led the section. On arrival, the guy with the best tracking ability took over and then the system worked on a relief basis. Swannie would fit himself into this order where it suited him on the day. He was not the best rider in the section so did not attempt to set the pace. He left that to the better riders, which was wise. As anyone who has ridden will know, it's easier to follow a fast rider than to compete, complex, but that's how it worked. If for some or other strange reason we got lost, Swannie would take control and appoint someone to show us the way, even when he was just as lost as us. Our system was good and it worked. Very few squabbles or ego problems, just fit in or get fucked up!

*Scramble...*SACC have hit the shit again. Same area, just further north. We left at speed, Lionel in front carrying the tracker, myself second, and Swannie third out the gate. I believe that Johan le Roux was behind Swannie, followed by the rest of the section. Within metres of leaving the base, Swannie overtook me and told me he was up behind Lionel. This was unusual to me because Swannie normally rode third, fourth or last in the formation. We took the route through the cucas heading for the bus route, we were bunched up and moving relatively fast. About 500m before hitting the junction going right to the Delta pipeline, Swannie gave me the *spread out/landmine* signal with his left hand. I slowed down as I was up his arse and in turn gave the signal to Johan, who was up my arse and trying to get past me. (He was a better rider).

As I looked forward again, I saw that Swannie wasn't going to make the right hand turn into the bus route that Lionel had made. He overshot by a few metres and got control into the corner on an old parallel track. I followed him and as I straightened out behind him, looked up…!

BANG…Swannie was gone, in his place was a dark red flame swallowed by a thick blackness. I was lying in the dirt, rifle at the ready trying to figure out what the fuck had happened. Swannie was gone…about 3m to the right of where he should have been was a naked povo spinning around in a circle, he was on the main tweespoor, the track we should have been on! My first thought was… AMBUSH, as I was about to shoot him I heard someone scream '… *moenie skiet nie, moenie skiet nie dis 'n myn!*'. A TM 58 had detonated not more than 15m in front of me and I could still hear someone shout, '…*don't shoot, it's a mine!*' Someone had taken control.

The poor povo, blood pouring from his nose and ears, dropped like a sack. Till today I don't think that cunt understands how close he came to dying twice in ten seconds. He happened to be bicycling himself in the opposite direction on the bus route when Swannie hit the mine not more than 3m from him. A lot of things happened pretty quickly after this, but it's worth pausing to reflect on the povo. His eardrums were burst and he looked like he had been completely sandblasted on his right-hand side. Obviously he was comatose, but fuck, the cunt was alive. Swannie was gone. 5/1/81

I think it was Lionel that went through the motions of the necessary drills, but by then we knew there was no further threat, and we consolidated the area then started to look for Swannie. It still hadn't quite sunk in that this had actually happened to us.

If my memory serves me correctly, the tracker with us that day was Sergeant Rooibaard. I say this because he was with us a week prior to the first SACC follow up. Some have stated that a tracker by the name of Kevin O' Neill was with us. Not so, if so then I'm wrong. I remember small details of that day. Like one of the guys calling the incident in, walking around the site looking for Swannie, and finding pieces of him. His two boot soles lay on both sides of the crater, as if he'd stood over the mine when it went off. Taking a local from a cuca close by and sticking his face into the biggest part we found, then walking away and muttering '…*one down and pop a wheelie*'. This is where Johan le Roux told me to shut the fuck up. One incident here comes to mind. Myself, CP Nell and (I THINK) Olwage were searching the surrounding area when we found a very small piece of Swannie. One of the guys said to me as I was about to scoop him up, '…*laat hom lê waar hy geval het*'. We covered a small piece of him with Ovamboland sand.

Some guys from whatever company was in the base at that time arrived, loaded Swannie up and did the necessary. We rode back to

The remains of Swanepoel's XR 500. Note bent R4 at left.

base and all credit to the base commander, were allowed to get pissed. Not even the dominee bugged us that night. I remember sitting with my roommate Simon Barker, drinking beer and trying to make an issue of the whole thing. There was no issue, we were alive and he was dead. The next day, 53 sent a Puma to pick up Swannie. We all grabbed a handle and walked him over the wall and placed him in the chopper, they took off and then he was gone.

Go out again? You are fucking insane!! Then this amazing loot came to our lines. I think he was the incumbent company 2IC, this is two days after Swannie got killed. He says, '...*manne, ons gaan op patrollie, ek gaan saam!*' This guy climbed on the BACK of a bike and took us on patrol. We patrolled the worst areas, rode on unswept roads, went back to the mine site, and he kept swopping bikes. Not riding, just pillion, we were fine after that. What a man.

We got pass, Johan and I lived in Vryheid at the time. He phones me one day and says, '...lets go to Pretoria, on the way I have to drop of some stuff for Swannie's parents.'

Fuck me. 'Ok.' Meeting Swannie's family was not on my agenda. We started with his ex-girlfriend in Carltonville because it was the only address we had. Eventually we find her place of work, this is where I started to realise that there's more to life than just ME. Johan and I walk into the bank, ask for...? This pretty girl comes out from the back office, looks at us and starts crying. She had never met us before but knew who we were, Swannie's mates.

Swannie's father was not at home when we got to the house, he was at work and there was no advance warning, but Swannie's mom was. The ex had phoned, so she was waiting in the driveway as we arrived. I learnt more about Swannie in one hour than I did in 18 months. Mr and Mrs Swanepoel's son did not die without being remembered. Fantastic family.

Johan and I left for Pretoria, we didn't say much till we got to his mate's house. I took an extended leave, then did an epic 22-hour drive back to Otavi (the unit had relocated) where I was given a 21-day suspended sentence for awol.

We carried on with the normal shit, follow-up after follow-up.

Swannie's grave at Fochville, near Potchefstroom.

40
Evel Knievel Se Moer

After the new XR 500s arrived, we spent many hours riding every day and we all became expert riders. A few of us raced motocross or enduros and used the unlimited free seat time to hone skills or learn new ones. Besides racing around the tracks we laid out everywhere, wheelying was one of our favourite pastimes. From lots of practice and some spilt blood, we learned to pull endless balance point wheelies at 80 plus in fifth gear.

We always rode in some kind of formation, even on one wheel. One day on a huge shona south of Ogongo we had fourteen okes spread out in a line, running through the gears on the rear wheel. We never tired of wheelying. After a while only a few bikes had speedos. They were flattened when we flipped, and removed.

Ovamboland is extremely flat. Not much opportunity for jumping, except over the bases of big anthills. To compensate, we did some really stupid shit on the tar road. It ran from Oshivelo gate all the way to Ruacana in a mostly straight line and we sometimes spent an hour or two travelling on it. That always got boring, so we invariably started fucking around.

One of our favourite games was riding in mass formation. Six okes would ride next to each other taking up the width of the road. Five then stuck their front wheels in between the rear wheels from behind, then four, three, two and finally one at the tail end. It formed a triangle of bikes, flat side forward, travelling at our usual tar road speed of 90 to 100. The sound of two dozen bikes that close, and I mean maybe half a metre apart, made the hair on your neck stand up.

It didn't take long before okes started kicking and shoving each other. When that got tame, you'd lean over and hit the killswitch on the bike next to you when the oke wasn't looking. That was always good for a few laughs. The formation exploded, riders jumped on the brakes, ran off the road or into each other. Great fun.

Playing chicken with oncoming traffic in the triangle formation was amusing too. I can remember a PB bakkie, one of those big white Fords with a canopy, taking to the dirt one day when we didn't budge. He went bouncing past on our left in a big dust cloud with the rear hatch flapping up and down, and we laughed our arses off. We played that game with army vehicles too, but they usually didn't chicken out so we had to. Even twenty bikes didn't impress a Buffel driver much, we learned.

Another fun trick we liked to pull on slow army vehicles was to sneak up on them from behind at high speed, split into two lines and go screaming by on both sides, hard on the throttle. The left line in the dirt kicked up a bunch of dust and combined with the noise, it usually scared the shit out of the driver. We got yelled and sworn at a lot, and never hung around afterwards.

Some days, usually after a lost argument with a Buffel or Samil had scared the shit out of us, we turned down the volume a bit and just cruised. We all carried sleeping bags and other kit on the back of the bikes, which made perfect backrests. By putting your legs over the handlebars and resting your feet on the front fender, you could cruise along as if sitting on a sofa. Only your throttle hand would be occupied.

On the tar road just west of Ombalantu, headed for Mahanene.

We never did figure out a way to lock the throttle in place so we could get *really* comfortable and enjoy the scenery. Maybe just as well, some okes would probably have fallen asleep. Not the time or place for a nap, doing 90 on a tar road.

41

The Tiffy

From time to time we grudgingly allowed outsiders to ride our XRs. During Ops Yahoo, we lent a bike to a Puma pilot and the show-off bastard promptly put on a wheelie display better than most of us could at the time. When we went to Tsintsabis in June, some camper Colonel rode with us one day. I forget who he was, but I can still picture him: a tall, dark-haired older oke with a greying moustache, a *poes beret* and glasses that turned dark in bright light. He looked like an older and bigger version of our enemy at Ogongo, the Intelligence Captain.

There definitely was a connection between those glasses and idiotic non-combatant officers. Judging by their behaviour, they apparently were issued with them once they proved capable of a certain level of stupidity.

This *Skiet Piet* Colonel got the bright idea one day that we could sweep the Bravo cutline between Tsintsabis and Oshivelo on our bikes every morning. We had ridden it a time or two, as expected we had to keep the speed up in the thick sand. The loot and I told him several times that it was a kak idea, we couldn't ride slow enough to pick up any spoor. He didn't believe us, so we invited him along for a ride.

It was a fucking riot.

We turned off the road onto the cutline and headed west towards Oshivelo. Out of sheer spite and to prove the point, we rode as slow as possible, just barely in control. We hadn't gone one click before the Skiet Piet *bliksemed* down in the sand. He dusted off his insulated PF bush jacket, straightened his *poes beret* and remounted. He didn't make it out of second gear before he ate sand again. After three falls in less than five minutes, he gave up and we rode back to base. The platoon was choking with the effort of not laughing out loud. As soon as we thought he was out of earshot, we cracked. Maybe that stupid bastard heard us laughing at him, because we were sent back to Spes the next day. We never set foot in Tsintsabis again.

On the Bravo cutline between Tsintsabis and Oshivelo.

Later, at Ogongo, we swapped rides with the berede platoon. They were our brothers in arms from Equestrian Centre and SWASPES so it was done in a spirit of comradeship, supposedly. In return for them riding our XRs around at the *asgat*, a few of us had the dubious pleasure of going on the dawn patrol around Ogongo on horseback. It was an hour's ride, one lap around the base, looking for anything suspicious. My ride lasted ten minutes. The four-legged, small-brained brute I perched on immediately sensed that I didn't know the first thing about riding him. He probably also knew that I didn't trust him because one of his distant cousins had kicked at me when I was eight.

Regardless of who started the shit, it ended with me donnering off the horse soon after leaving the gate. I walked back to camp, picking up my bush hat where it had blown off my head because both my hands were locked in a death grip on the saddle horn. While the horse was galloping out of control and just before I impacted Ovamboland, that was. Those bastard donkey drivers rode off laughing at me. That was the last time I ever rode anything with its own brain.

A chance meeting with berede platoon 28 on the Dombondola shona.

During our time at Mahanene, a 3 SAI tiffy whom the troops befriended wanted to go ride with us. He was a tall, skinny oke with dark hair and a thin moustache and looked like any one of us. Except for the pair of greasy short-sleeved overalls and web belt he lived in that shouted *TIFFY!!* loud and proud. He had won his class in the Roof of Africa Rally in Lesotho the year before and was obviously a tough-as-nails, expert rider.

One of the troops loaned the tiffy his bike for a patrol in Angola. Some time on the third day we rode back to Mahanene to get petrol and when we left, our mate was back on his bike. The tiffy was missing in action. I never saw him again. The troops told me he said we were '...*fucking crazy*'.

It would seem that chasing terrs all day and sleeping in TBs in Angola was not his cup of tea. I was surprised, but I could see his side of it. When you think about it, it probably was a bit more dangerous than racing the toughest off-road race in Africa.

42
Pisscats And Sparks

We had three forms of recreation in the army, especially in the bush: darts, snooker, and drinking like fish. At SWASPES we honed those skills to a fine edge. In Ovamboland, we made the most of the few opportunities we had to get *vrot*.

Mahanene, occupied by a 3 SAI company, was the main outlet for our frustrations and tensions because we spent six or seven nights there. Captain Rocha, the ex-Angolan company commander, wore shorts and didn't have a saak with SADF rondfok. Unfortunately for everybody, his CSM was a mental case named Sparks. Sparks was a PF temporary Staff-sergeant, a big paunchy lout with curly black hair, a big nose and a weak chin. He had a bad reputation back in Potch where 3 SAI and Berede were based. I knew him through the grapevine and from occasional previous encounters. He was a pisscat and troublemaker on semi-permanent extra duty. His fellow NCOs at 3 SAI tolerated him for that reason only, I'd been told.

One quiet evening at Mahanene we sat around a big fire talking kak. A 3 SAI ex-Rhodesian platoon sergeant named Boris entertained us with his dry sense of humour and stories told in the third person. He was a lanky oke with sandy hair who had fought in Rhodesia for years, so he felt fuckall. Boris ran out of tall tales and fell silent after a while. For further amusement, a captured swap was brought out of the POW cage to sit at the fire with us.

Sitting around a fire with one of the enemy and drinking beer was a disorientating experience. He was our age. In fluent Afrikaans he answered our curious questions as if we were old pals. He told us that they were very afraid of the bikes because we appeared so suddenly out of nowhere. At one point my loot got annoyed and lunged across the fire at him. I forget exactly why, but it was over something he said, some SWAPO propaganda he'd been taught. I calmed the short-tempered loot down before it got out of hand.

The conflicting emotions of the situation baffled me. Like many

115

of my peers, I was very aggressive with a trained Pavlovian violent response to anything SWAPO. Under different circumstances we would have killed him in a second, and vice versa. I remember staring at him across the fire, confused by how much he was like us and almost liking him. It didn't fit into the template of the *Rooi Gevaar,* according to which I had to hate him and kill him.

We only spent about a week in total at Mahanene, but got more grief in that short time than at all the other bases combined. All of it was Sparks' doing. He held a fucking parade early every morning and always found some reason to shit on bike squad. One morning he was completely beside himself. He was still pissed from the night before, his hangover hadn't even kicked in yet. With eyes like pissholes in snow and alcoholic red face, he stood on a low podium ranting and raving at us.

More specifically at *me,* three metres away.

The brown paint on our bikes was rubbed and scratched off from daily wear and tear. We also scratched girlfriends' names, nicknames, and slogans into the paint. Between 1980 and 82, the big Japanese motorcycle manufacturers produced the first dirtbikes with single-shock suspension. They gave their revolutionary new suspension systems catchy names. Honda's was called 'ProLink', Yamaha's 'Monoshock', Kawasaki's 'Unitrack' and Suzuki's 'Full Floater'. I owned a Suzuki RM 465 at the time, so brand loyalty made me scratch *Full Fucker* into the paint on both sides of my bike's swingarm to make it look like my RM back home.

Sparks had seen this while *sluiping* around the base that morning and he was coming unglued. He stood on his little elevated box, red-faced, eyes bulging, and yelled and screamed about our '...*undisciplined behaviour...damage to army property...communist this...un-Christian that bla-bla-bla....*' Those were the highlights I remember, the rest was just noise. He was snot-flying angry over FULL FUCKER that he had seen on one bike. When he asked whose it was and I raised my hand, he nearly blew his ring gasket.

The spectacle of a self-righteous, degenerate alcoholic calling us every name in the book was offensive and hilarious at the same time. I stood there staring at him, debating whether to shoot the cunt or collapse laughing.

The chaplain in Mahanene at the time was a conniving twat who clearly wished he were an infantry corporal instead of a man of the cloth. He tended to forget his position in life and tried to fuck us around at every opportunity, but we lagged him af. Not long after the noisy parade, he and Sparks ganged up and threw us out of Mahanene. We spent the rest of our time sleeping in the Angolan bush, but it was more than worth it. The monumental piss-up that finally caused our expulsion was the kind you tell your grandchildren about when they climb up on your knee and ask: '...*grandpa, what did you do in the war?*'

Fast forward to 1983. I ran into Sparks in the Potch NCO mess at lunch one day. The army had a fucked up system of temporarily promoting NCOs in the bush for chain-of-command reasons, and he was back to his permanent rank of Sergeant. The beauty of it was, I had been promoted to Sergeant right after returning from Ovamboland. I went over to his table and asked him if he remembered me. That dumb bastard did not. I reminded him about Mahanene and it got ugly quickly. I forget what I said and did, but I had enough sense to walk away when things turned violent. A fist fight in the mess over lunch would have landed me in the clink or got me demoted. I didn't want to be like him.

That was the last time I saw Sparks. Eight or ten years later, one of my childhood friends who had done his national service in 3 SAI told me that Sparks had been hanged for murdering his wife and the oke he caught her in bed with. I was not the least bit surprised.

43
Pyromania

Besides not *trapping* around with 40 kilos of shit on our backs, we had another advantage over the SAI *bok-kops*. We could make a fire anytime, anywhere, regardless of the weather. That came in very handy once it started raining. We'd gather a pile of wood, douse it with a litre of petrol siphoned out of the bikes, throw a match and have a roaring fire going in the middle of an African thunderstorm in five minutes flat. It saved us from much misery after the first night we got wet in Angola, and on a few occasions after that.

A bike squad fire is an easy way to get a braai going, but it's not for amateurs. I have learned that, unlike in the bush, petrol practically explodes in an enclosed braai and standing too close when you throw that match is not a clever move.

Bike Squad braai fire in Angola.

44
Noddy Cars

As part of the campaign to keep us biker scum out of their bases, the PFs in 52 Bn attached us to an armoured car squadron for a week. It was quite an enlightening time. Those okes were *very* serious about going to war as comfortably as possible.

Operating with the Noddys.

Neil, the two-pip loot noddy commander, happened to be a childhood friend of my loot's so it was an amiable and fun week. They shared their overabundance of food and drink with us. We couldn't offer them more than half a tin of bully beef or sweetcorn in return, but they didn't seem to mind. We lived on ratpacks, but their flatbed Buffel log vehicle carried boxes and boxes of food and drink. They had it all: fruit salad, Ultramel custard, *ovambo piele*, corn flakes, rusks, and lots of those ratpack milkshakes everybody fought over. Or traded for disproportionate amounts of mixed veg or corned beef hash when you got really hungry. They even let us put our kit on the flatbed Buffel, which made for thoroughly enjoyable riding. The oke who'd been carrying the B25 radio was especially happy.

Our bikes were much more mobile than the noddy cars, so we roamed away from them and RV'ed every so often. Crossing a shona just south of the cutline one day, the bikes were off on the left flank moving at a good rate of knots when the column suddenly veered sharply right. I was at the end of the line as usual and the whiplash effect had me turning hard in a power slide, trying to stay in line and not slingshot into the bush. By the time I got up front, the action was over. The noddy in the lead had seen three okes carrying AKs leisurely riding south into Ovamboland on two bicycles. It was touch and go for a few seconds. Turns out they were UNITA.

The noddy squadron's shit-stirrer was a naughty Porra from the south of Joburg, so communication with the three Angolans was a piece of cake. Somebody pulled out a camera and one UNITA reacted by posing in heroic stances he must have seen on old communist propaganda posters. I cracked when he did the 'cockroach': on his back, legs up like a dead insect with knees bent and ankles crossed, AK in *hoog voor*. I snapped a photo a second or two before he did that and missed a classic moment. We gave them a few grenades and water bottles before parting as friends.

Our UNITA friends acting war-like. Hagen standing at right.

A few days later we ended up in Oshakati. The loot and I went on a noddy night security patrol through town. It was a boring few hours, driving aimlessly through the quiet base and surrounding town. The only excitement occurred when 81 mil mortars and the 40 mm cannon on top of the water tower shot a fire plan into the countryside, supposedly to keep any swaps out there on their toes. I wondered how many donkeys and pigs got their arses shot off that way, wandering around freely like they did.

Another memorable occasion was a conversation a dozen or so of us had with a PB late one afternoon. He was obviously the big man in the neighbourhood, and had been klapping the local brew of choice since at least noon. We had an hour of fun with him, talking kak and swapping stories.

The noddy crews were good okes, but we simply could not relate to their luxurious lifestyle. They washed and *shaved*, for fuck's sake. They in turn thought we were animals, carrying only sleeping bags and a few tins of food. I realised they were in a different army than us when I saw Neil and his platoon sergeant set up camp beds with fluffy white pillows the first night out. I couldn't believe my eyes.

Sleeping in a bed in a TB. Jissis…where the fuck have you ever heard anything like that in your life?

Kakpraat session with a talkative PB.

45
Brothers

The first rains of the season caught us somewhere in Angola. We'd had a typical day, charging around in a foreign country looking for swaps to kill. Nothing stands out in my memory until some time in the afternoon, when we found a big dam and took a siesta. We were hungry so I went fishing with a hand grenade. Two dozen smallish dead fish floated to the surface. Visions of fresh fish braaing on an open fire were running through our brains, and I appointed two volunteers who stripped down and waded into the dam to collect our dinner. We made fires and braaied the fish, but to our great disappointment they tasted very muddy. We chucked most of it away and went hungry instead.

I had taken my boots off and rolled up my pants, and was standing knee-deep in the dam when some clown tackled me from behind and knocked me down. I was soaked and took my clothes off to dry. Soon after, the loot got antsy and decided to move out. My clothes were still very wet so I tied it to the back of my bike and put my boots, webbing and R4 on. I brought up the rear of the line completely *kaalgat*. Apparently I made a strange sight. The troops were laughing at me, and I know one took a photo. I do wonder what would have happened if we'd hit a contact then. SWAPO probably would have laughed themselves to death at the naked lunatic on a brown motorcycle with a rifle on his back and clothes flapping behind like ragged flags.

The weather became a serious concern about the time my clothes were dry enough to put back on. Huge thunderclouds were building and we tried to outrun them, but it was too late. We were surrounded by dark, threatening skies. Lightning bolts, loud thunder and gusty winds made it clear that we were about to get rained on big time. In premature darkness we hurriedly picked a spot for the night's TB.

Unbeknownst to some of us, we chose a slight depression to pitch our hopelessly inadequate bivvys in. Our sleeping bags were made of synthetic material, the bottom half was 'water resistant' and the top was a nylon/cotton blend. It started raining with a vengeance. About the same time I started feeling intimately connected with nature, I realised that the pressure against the bottom of my sleeping bag was water. I was in denial for a while, hoping that it wouldn't get deep enough to flow into the bag. No such luck. By the time the water was 10 cm deep I knew I was in for a miserable night. So were a few other okes in my vicinity including Koos, the loot.

All I remember about the rest of that endless night is that I was very wet and got colder and colder. Make no mistake, Africa can get bloody cold if it's nighttime and you're soaked through, even in summer.

By first light several of us were in the initial stages of hypothermia. We were shivering uncontrollably and knew we were in trouble. The sky was cloudy and no sun would come out to warm us. The okes who had been lucky enough to sleep on higher ground took care of the drowned rats. The situation got so serious that Sakkie, one of the section leaders, made the loot and me crawl into his dry sleeping bag dressed only in our rods. Neither of us had any reservations about it, all we cared about was warming up and being functional again.

After an hour or so we were over the worst of it. We made a huge fire and gathered around it in various stages of undress to dry out and warm up. We didn't particularly give a shit about SWAPO or the war at that point. As it warmed up our spirits rose with the temperature, and soon somebody laughed for the first time that day. The TB looked like a squatter camp with clothes, sleeping bags, bivvys and webbings hung over bushes to dry. The sun finally made its hesitant appearance, accompanied by the frog chorus. We moved on in the late morning after our clothing and kit had dried out.

The next year, back at Berede in Potch, I got called on the carpet by the RSM. I was way too chummy with the officers, he said. He called it 'verbroedering met offisiere'. I was fucking incredulous. Herbert and Kobus, the two loots I had spent a year in the bush with, were my close friends. We were all in our third, and in my case fourth, year of short service. We maintained discipline and professionalism when needed but when alone or off duty, rank and formality went out the window. We were 20 and 21 years old for fuck's sake, and had been together in many situations where only pure luck kept us from being killed or badly hurt.

I told the RSM the story of me and the loot crawling into one sleeping bag in Angola to recover from hypothermia, and asked him if he honestly expected us to forget about it and live under rank apartheid. Also, that off duty I would associate with anybody I wanted, it was none of the army's business. It turned into a bit of a stand-off. I always thought he had a personal grudge against me and braced myself for extras or worse. To my great astonishment, he seemed to understand and merely told me to fuck off.

I was especially good friends with Kobus, the loot I went to Ovambo with. We lived together under harsh conditions for months, drank together, went to motorcycle races together, pomped university pundas together, and were generally like brothers. I haven't seen or heard from him since the day we klaared out. We didn't even exchange contact information. I suppose at the time we were just too focused on the klaaring out process, and we left the army on different days.

It has bothered me for years. Despite many efforts to track him down, I still have no idea where he is or if he is even alive. I hope to see him again some day soon. Our business here isn't done yet, we have much to catch up on.

Dinner near Ogandjera, a can of bully beef or mix veg stuck in the fire.

46
The Party

Mahanene was a small base near Ruacana, four kays from the cutline. It was next to a fish farm and had a few permanent buildings, a swimming pool, and the nicest bar I ever saw in Ovamboland. Despite the amenities it was a tense place. Everybody seemed on edge, whether from its reputation for getting revved with mortars or the volatile 3 SAI Company Sergeant-Major who ran the base, I could never figure out. Either way, the tension resulted in manic behaviour by the occupants at times.

The round pub had a wooden fence around it like an Ovambo kraal. To the right of the entrance was the bar counter under a small thatched roof. Beyond that was a huge braai flanked by benches, built with cement and rocks imported from far away. To the left of the centre pole which held up the shadownet roof was a fishpond, and along the left wall beyond that a snooker table. Between the braai and the snooker table a new dartboard hung on the wall.

The pub was destroyed one evening during the wildest piss-up I had the privilege of partaking in my entire time in the army.

We rode into base around three o'clock on that day, straight out of Angola. Because Mahanene was so close to the cutline it was a convenient fuel and booze stop when we operated in Angola. The loot and I jaaged right up to the pub entrance. We parked our bikes on opposite sides of the path and walked inside with helmets, webbings and R4s on. We were hot, dusty and very thirsty. I downed a cold beer, followed by three big glasses of Bailey's and milk that produced some very impressive projectile vomiting.

After that, things got blurry. A dozen or so people were in the pub at any given time and things got out of hand, as usual. I remember throwing okes into the fishpond and later the swimming pool. One loot who had gone to bed was dragged out of his room by four drunks. We had to pry his hands and feet off the doorjamb to get him outside. He fought like a tiger, but he went for a midnight dip. Next thing I remember is waking up in one of my platoon's tents, behind a trommel under a bed in a corner. It was mid-morning and I had an absolute *cunt* of a hangover. I stumbled around for a while in the strangely quiet base. No rank was to be seen anywhere, and I eventually wandered into the pub looking for survivors.

The beautiful pub looked like a mortar had landed in it. The thatch roof over the bar was pulled down. Left-over food, paper plates, empty beer cans and broken glasses lay everywhere. The snooker table was upside down, the broken balls and sticks nearby. The dartboard was folded in two. Everything was covered in a layer of white powder from empty fire extinguishers scattered around. The image I will never forget, is the two little goldfish cowering under a rock in the fishpond which had beer cans and a half-eaten T-bone steak floating in it.

As I walked out cracking myself, the chaplain ambushed me. We had clashed with him before. He liked to throw his two-pip rank around but we figured outside of church, a *siel tiffie* had no authority over us so we lagged him af. He was very disgruntled with the heathen infantry. I suppose he picked on me because he happened to bump into me, but I had the *babalas* from hell and didn't have a saak with him. My throbbing head was in no shape to listen to judgmental rants so I just walked away.

Within a day or two our bike platoon was kicked out of Mahanene. Courtesy of that cunt chaplain and Sparks, the CSM. We spent the rest of our time sleeping in TBs in Ovamboland and Angola. It was definitely worth it. That piss-up set the bar very high. I have never been in one even *close* to that wild since, and not for lack of trying. Problem is, people in civvy street don't appreciate the sheer destructive joy of it. In the fucking police state I live in at the moment, you'll get thrown in jail if you pulled a stunt like that.

On the website Google Earth, Mahanene is clearly visible. The shapes of the pub and swimming pool can still be seen, even from space. I will go back there on a bike soon. Call it a pilgrimage.

47
The Frogs Of War

We were some distance inside Angola when the first rains of the season came. We didn't TB till after dark, which turned out to be a bad mistake. Some of us unknowingly picked a slight depression to sleep in and got very wet that night. When daylight finally came we made a big fire, something bike squad could do rain or shine. All thought of SWAPO and being in a hostile foreign country forgotten, we stood around the fire and hung our wet clothes and sleeping bags on bushes to dry. The landscape was transformed overnight by the rain and already halfway green. The warming sun soon restored our appetite for making kak, and okes were laughing and carrying on while we made coffee and got fully dressed again.

After sunrise we heard a very strange and sinister *wup-wup-wup-wup* sound coming out of the bush. It sounded like that old video game *Pac-Man*, but much louder and from a thousand different sources all around us. We couldn't identify it and got a bit jumpy, being only 25 okes in the area rotten with SWAPO bases where Operation *Smokeshell* had taken place not long before. I remember thinking it was a bit what the Boers or Brits must have felt like, surrounded by 15 000 Zulus. Eventually someone discovered a big frog merrily croaking away and the mystery was solved. Soon we found masses of frogs all around that came out of hiding after the rain. They were obviously happy and did not allow their limited two-note repertoire to detract from the celebrations.

But of course it didn't end there. A commotion started some distance away, yelling, swearing and laughter. I walked over to see what was going on and found six or eight troops surrounding a frog sitting in a ratpack box. What had this bunch excited was this huge frog, almost the size of a rugby ball and various shades of green with red spots, jumping and snapping at the sticks they were poking it with. It sported a set of choppers that would make a fucking piranha proud and was extremely aggressive. Eventually someone got bitten, got angry, unslung his R4 and shot the frog on automatic. It exploded, which sent the onlookers into hysterics. Within two minutes it sounded like *Smokeshell* all over again as the whole platoon went frog hunting.

After inflicting heavy casualties, we eventually got bored with it and went back to drying out and moving on. We suddenly realised that every swap or Cuban within 20 kays now knew our exact position, and we were relieved to get out of the area. It was our last month in the bush and we were all getting a bit *bossies*, but we still had a healthy respect for what could happen.

In hindsight, it seems that every living thing in Angola except some PBs were hostile towards us. I did some research later and learned that we had tangled with an overwhelming force of African Bullfrogs, described as '*aggressive with sharp teeth*'. You don't say. If those fucking frogs had been any bigger, SWAPO would have been the least of our worries.

Mounting up after a relaxing braai in Angola. Corporal Fourie in left foreground, Steyn the Wheelie King far left.

48
Kamikaze Mouse

Within a day or two of me spooning in a sleeping bag with another man, we got caught by another afternoon thunderstorm. We had crossed back into Ovamboland and in vain tried to outrun the storm again. The cold night in Angola had taught us a lesson, so we stopped and TB'ed as soon as we realised we were in for another soaking.

Having also learnt the lesson about sleeping on high ground the hard way, we picked one of our usual hollows among the trees and pitched our bivvies on the high spots. The loot and I were perched on a small mound which stuck up high enough that it would be the last bit of dry ground in Ovamboland. We managed to clip our groundsheets together to make a two-man tent and roll out our sleeping bags under it before we got wet. The rest of the platoon did the same.

We were all soaked to the skin and it was still mid-afternoon, so we made a bike squad fire under a huge tree. It was a surreal scene: a bunch of wet okes standing around a three-metre fire under a tree while it pissed cats and dogs. We were happy though, we knew we would sleep warm and dry that night.

It was a major downpour. The hollows around us started to fill up and a few okes had to move their bikes to keep them from being drowned. We could visualise what the area would look like by morning. Small animals were scurrying around, chased out of their holes by water. A small mouse with two white stripes down his grey back came running between me and the oke standing next to me by the roaring fire. It was soaked and shivering and looked very sorry for itself. Obviously attracted to the heat, it walked up so close to the fire that its hair started steaming. I bent down, grabbed it by the tail and pulled it away from the intense heat. A few of us were watching it and were amazed by its lack of common sense.

The mouse sat there for maybe ten seconds, looking at the white-hot interior of the fire through a gap between two logs. I felt I had done my good deed for the day and expected Mr. Mouse to dry himself and fuck off home. Instead he flicked his tail, twitched his nose, and walked right into the fire through the gap between the logs. He simply disappeared. We cracked ourselves. We didn't think that animals could be nearly as stupid as humans, but this one was. Of all the strange animal encounters I had that year, the kamikaze mouse was the most bizarre.

When it got dark we went to bed. It was still raining like a two-cunted cow pissing on a flat rock, but we slept dry and well. I took off my wet clothes and tossed it next to me. I was warm and dry and so was my R4. I didn't have a saak with SWAPO or the world. Some time during the night it stopped raining, and when the sun came up we were on a different planet. I was lying in a little tent on a small island. All around me were brown bivvies on mounds sticking up out of a small lake. The hollow had filled with water and the landscape was transformed.

It took a while to get organised and move on. Some okes had to wade chest deep to get to dry land, and close to me a bike stood half-submerged. The rear wheel was almost completely under water. The oke who carried the B25 radio had left it on the ground next to his bike. He found it by stumbling over it. When he picked the radio up, water poured out of it. We looked at it and laughed. We had *min days*, we had slept nice and dry, and we felt fuckall.

We left Ovamboland two weeks after that. I never thought I could relate to a rodent, but that suicidal mouse still puzzles me sometimes. After the miserable night in Angola, I knew exactly how he felt. I didn't walk into a fire like he did, but maybe that was a metaphor for life since then, who knows?

49
Min Days

By November 1982 we'd been living in the bush for months. We were like redheaded stepchildren, no base wanted us. Or rather, PFs at every base in 52 Bn didn't. We'd been thrown out of three up to that point, so we logically assumed we'd be bush dwellers for the rest of our time in Ovambo and Angola. We were all getting a bit *mal* by this point. The games we played got more and more dangerous and we were very *nafi*. We felt the PFs were sending us on bullshit patrols just to keep us out of their hair.

One four-day trip into Angola was clearly that: ride to point A and report in, ride to point B and report in, ride to point C etceteraa. So we found a nice spot and laid out a MX track instead. We stayed there for two days, racing around like madmen and having a jol. Being about 50 kays inside Angola, we had to calculate our fuel very accurately. When we had barely enough left we rode to the nearest base, filled up, and went right back to Angola. We made another track and spent two more days happily jaaging around. It was on one of those days that I laid out a figure 8 around two bushes, and went round and round dragging the handlebars and footpegs in the damp sand. Unfortunately that dirtbike gold medal stunt went undocumented because we had all run out of film by then.

The loot making the daily SITREP from Angola on the bulky B25 HF radio.

The loot called in the required *sitrep* every evening,

with the coordinates of the points we were supposed to be at.
Our attitude was, fuck 'em if they can't take a joke. Or give us a

worthwhile job. At that stage none of us wanted to deal with the kak that accompanied life in base like sweeping the sand, chicken parades, *klaarstaan* every morning, or running 2,4s because some PF thought we were *nafi*. Neither were we interested in dealing with the wankers we invariably clashed with, like the oxymoronic Intelligence Captain at Ogongo or the unhinged CSM at Mahanene. We were much happier out in the bush.

Somewhere north-east of Mahanene one day we stopped under some big trees for our midday siesta. The body of an old Ford bakkie lay nearby. In our endless quest for thrills and amusement, we loaded two bikes on the back and Van and Haps climbed inside the cab. The plan was to take photos and send them to Bike SA with the caption '*SADF Roof of Africa team*'. To add some atmosphere to the photos we were taking, I chucked a smoke grenade under the bonnet. The red smoke curling out definitely added colour to the scene.

As usually happened, things soon got out of hand. One oke decided the scene was still too dull and threw several grenades at once. It worked, all right. The chemical smoke filled the cab through the rusted-out firewall and we could hear the two okes inside coughing and swearing. Next thing we knew they bailed out through the front window frame, choking and eyes streaming tears. That got the usual reaction from the platoon: laughter, jeers and insults.

Soon after this game ended, someone noticed movement on a branch above some okes lying in the shade. Closer investigation revealed two long and skinny lime green snakes. Brilliant! Now we had a pair of deadly two metre long *boomslangs* to play with. We tried to get them out of the tree with sticks, but they moved higher and out of reach. Someone unslung his R4 and shot one snake, after several shots it fell to the ground cut in two. We were all curious and crowded around fascinated by this insignificant-looking thin snake, one of the most lethal in Africa. Until some junior herpetologist came up with the story that *boomslangs*, like swans, mate for life and they avenge each other's death. That set off a minor contact as the other snake, high in the tree, was shot down. The idea of vengeful snakes that could kill with one bite made everybody jumpy so we threw them in the fire and moved out from under that tree.

Looking for more entertainment, I noticed two anthills nearby that were spaced just far enough apart to make a relatively safe double jump. I planned on racing motocross back in SA the next year so it was an ideal opportunity to practice under controlled conditions. I first jumped off to the side to judge the distance. After two or three attempts I had it figured out, and spent the next fifteen minutes happily doing doubles until it got boring. We were still wary of tree snakes so we moved on soon afterwards.

After the rains began, billions of mopani flies tried to drive us insane. If not for the invention of smoke they would have. East of Mahanene, November 1982.

Our last days went by uneventfully. Meaning, nobody got hurt or killed. In the middle of December we received orders to move back to SWASPES and then to SA. We were driven back to Spes in Buffels, the bikes on Kwêvoëls. The only thing I remember about the trip was having to disturb the usually relaxed atmosphere in the platoon when Luwes, who had torn his windpipe crashing through a wire fence, rejoined us for the trip home. I did a final check to ensure everybody was strapped into their seats for the five-hour trip on the tar road, and found him sitting in the bed of a Buffel by one of his mates' feet.

Buffels were excellent vehicles in the bush, but on roads they could be lethal to the occupants. The V-shaped body on a Unimog chassis was top-heavy and many men were killed by rolling Buffels. You were supposed to be strapped in at all times, but that was lagged af. Seb from platoon 11, who had logged the first crash on the new XRs, had an extremely close call when a Buffel he was standing up in rolled, but missed him. He got dumped out when it went upside down and it scared the fuck out of him.

Remembering that, I ordered Luwes to go strap into his seat on another Buffel. His connection Burley, who had disappeared from the platoon some months before but popped up when it was time to go home, apparently forgot about Seb's thrill ride and started shouting the odds. The argument escalated until I pointed out that I didn't particularly want to tell Luwes' mother that he died in an accident *on the way home*. That got through to him, and he eventually piped down. Luwes took his seat in the other Buffel and strapped in tight. Some okes never quite understood that the most important part of me and the loot's job was to get them home alive.

After a few days klaaring out at Spes, we flew back to Waterkloof in a Flossie. Before we left Otavi the MPs inspected everybody's kit for contraband. All ranks had to lay out their balsaks with all their kit next to it. It looked like a flea market where everybody sold the same shit. I had a few souvenirs, including some shiny 20mm HE rounds I got from the 61 Mech Ratel okes. I was determined not to give them up. Some devious mind figured out that we could ship things back to SA by rail. Accordingly, the trommel I acquired to transport my spoils of war was shipped from Otavi station right under the MPs' noses. A matter of principle, we despised the fucking Meat Pies.

The flight was uneventful. I remember nothing of it except the throng of people in the terminal at Waterkloof after we landed. One final clash with the RSM at Berede over my hair that was too long for his liking, and I drove off on a month's leave. I dropped *Six foot four and full of muscle* Dave at Joburg station along the way.

It was over just like that, the end of a life-changing adventure. Twenty-five years later I can tell when the habits it left me with surface, and most of the memories make me laugh. I'm beyond the point of wondering what it was all about and frankly, don't care anymore. Fuck it. Shit happens.

As far as I'm concerned, it was to create the muscle memory I enjoy every time I ride my pristine 1982 XR 500. It's a vintage bike now.

Somewhere in Angola

50
Jingle Bells

At SWASPES I ran into another Infantry School acquaintance of mine, a PF loot named Kevin. He was a friendly oke, dark-haired, short and stocky with the wispy moustache that was the best most 20-year-olds could muster. I vaguely recall that he was the SWASPES Adjudant at the time, which classified him as a jam stealer. I had the impression Kevin felt left out, as he always acted like a lonely Labrador around the two bike platoons.

He came from a military family. His *toppie* was a Sergeant-major at 2 SAI in Walvis Bay. They were 'Southwesters' and not South Africans, as he told us several times every day. Kevin had wangled a posting straight to Spes after getting rank because he raced motocross in SWA. He owned a brand new Suzuki RM 250, the first water cooled model. We all lusted after that yellow beauty but he never allowed anyone to ride it.

We didn't see Kevin around much. It seemed he was always off racing somewhere or visiting his large family. We never figured out whether that was with higher-up permission, or simply skillful gyppoing. The day in December when we arrived back at SWASPES from Ovamboland, he was there though.

I jumped out of the Buffel, a good two-metre drop even with the sides lowered. Kevin happened to be standing right next to where I crouched from the impact with earth, wearing a full webbing and R4 in hand. While I was still airborne, he started telling me that he had won some or other motocross championship. He must have thought that '...*hello, welcome back, glad you're in one piece*' was inappropriate. I looked up at him and said '...*that's nice, where's the beer?*' but his brain was stuck in 'transmit' and he rattled on and on about his budding racing career. I just shook my head and walked away.

Over the next few days we returned our rifles to the weapons store and klaared out to go back to SA. In between, we swapped stories. The other bike platoon had had some hair-raising experiences too.

You could feel the tension ease as we got vrot, talked kak and blew off steam. Kevin's main contribution was a family Christmas tale, seeing that it was a week away. He told us they didn't say Merry Christmas to each other, but '...*merry syphilis and a happy gonorrhea*' which they'd do over and over and piss themselves laughing. He said his mother and grandmother found it especially amusing. I just stared at him. I couldn't think of a snappy comeback for that one.

Kevin has been dead for over twenty years now, from a car accident two or three years after I last saw him. He may have had a strange family, but he was a good oke. Every Christmas I still catch somebody with his bit of ancestral wisdom. You have to be careful who you say it to though.

Civvy dirtbikes at the school in Otavi. Herbert on the left tying down his Honda CR 480, Kevin next to him with his new Suzuki RM 250.

51
Your Turn Is Your Turn Too

Two bike and two horse platoons returned to SWASPES in mid-December before travelling back to SA. We spent several days catching up with our mates from platoon 11, swapping stories and comparing notes. They'd had their share of close calls with landmines too. Herbert the German loot and a section *klapped* a double cheese mine in a Buffel one day, but they were all strapped in and nobody got hurt worse than bruises and burst eardrums.

That didn't impress us as much as Kruger's gold medal stunt though. The same Kruger, shit-stirrer and scrounger *extraordinaire*, who had been attacked by the baby baboon in the Ratel nine months before. Crossing the cutline one fine day, he rode over the edge of a cheese mine and popped it up out of the sand like a frisbee on a beach.

It doesn't get much closer than that. Herbert told me that Kruger was different after that day. I'd always thought of him as a future jailbird, but that may have been a pivotal moment in his life.

Relaxing at SWASPES after returning from Ovamboland, a week before Christmas 1982. Berede loots Nic and Martin at left, my replacement as Platoon 11 Pl. Sgt. at right.

52
War Pigs

One of the great ironies of the bush war was the fact that we were the first South Africans to live in a racially integrated environment. Discrimination still existed, for instance there were no black officers except in 32 Bn, but life 'on the border' was fundamentally different from civilian society at the time. Officers, NCOs and troops of all races lived together, ate together, showered together and fought together. And more than a few died together.

I read statistics many years later that more than 50% and as much as 70% of the army (SADF and SWATF) was non-white. The classic example was 32 Battalion, made up of black Angolans and mostly white SA officers and NCOs. We had no contact with them, so I have no insight into their daily lives.

We did spend quite a bit of time with Koevoet, and had some interaction with 101 Battalion. My Infantry School platoon in 1980 was attached to Koevoet for several weeks and we slept in their bases, drank in their pubs, and played on their volleyball courts while operating with them. In early 1982 we spent many long, tense days with Koevoet, chasing terrs around the bush during the autumn farm olympics.

Nowadays the self-appointed experts who were never there slander their reputation, but it's mostly revisionist kak. Undoubtedly some people got donnered by mistake and a few died the same way, but in my experience Koevoet was fucking brilliant. They were not the undisciplined killers committing atrocities everywhere they went, as some wankers try to portray them. They were the best counter-insurgency force ever, pure and simple. There was no rondfok. They just lived to kill swaps. I respected them then, and still do, immensely.

Koevoet was unconventional by army standards and many PFs hated them for that. But I suspect it was also because they got 90% of the kills in Ovamboland and made the army's name *poes*. Their vehicles always looked like adult toddlers' rooms. Ratpacks, sleeping bags, clothing, weapons, ammo boxes and spare magazines were scattered all over. I remember riding in a Casspir with RPG-7 rockets and rifle grenades lying around like discarded toys. The Casspir wasn't spacious, so you had to watch where you put your feet.

I never saw any saluting or visible rank and never heard any yelling while operating with Koevoet. Those skinny constables could track like bloodhounds and run like greyhounds. Some of them were former swaps who had been reorientated and shown the error of their ways. Their intelligence gathering capability was incredible. They were locals, spoke the language, and often knew the people they encountered. Personally, I always liked working with them because action and results were guaranteed when Koevoet was around.

We dealt with PBs on a regular basis. PBs were the *plaaslike bevolking*, meaning the civilian population in Ovamboland. In Angola they were known as *povos*. As in any war they were the ones who got the short end of the stick, especially in Angola. Our relations with them were mostly friendly but formal. We were not trained to gather intelligence so the extent of our questioning was usually *'... where is SWAPO?'* which didn't help much.

When we got flats, we usually borrowed a pump from the nearest kraal. We routinely relied on local knowledge to help us navigate the flat, featureless terrain. We soon learned that *'not far'* or *'just there'* while waving and snapping fingers in a general direction could mean two kays or twenty, so we structured our questions accordingly. Kids were always fascinated with the bikes and we had hours of fun giving them rides. We handed out lots of ratpack sweets, and aspirin and bandages once in a while. People in Angola would flag us down and ask for medical help. As sorry as we felt for them, there wasn't much we could do. It always surprised us that we met few PBs who didn't speak Afrikaans, even deep in Angola. It was another of the paradoxes that contributed to the conflicting attitudes we had about the war.

No conflicting feelings about SWAPO though. Infantry leadership at platoon level consisted of selected, aggressive and very well-trained individuals. We were very keen to kill swaps. I never once felt the slightest fear of them, even when facing their elite forces during Ops

Yahoo. They were the faceless enemy. The night at Mahanene when the only live terr I ever spoke to told us in perfect Afrikaans that they were shit scared of the bikes because of how suddenly we appeared, was a bit confusing. We had been trained and conditioned to kill anything SWAPO. The emotions I felt, sitting around a fire with an enemy who suddenly became human to me, ranged from liking him to extreme aggression.

Today I still have illogical, contradictory feelings about the war and the former enemy. I bear them no animosity. I don't think I do, anyway. Sitting down over a few beers with a former SWAPO cadre and exchanging stories about our experiences would be very enlightening. I suspect we'd get along well, and I'm glad it's all in the distant past. But part of me is stuck in 1982, in a parallel universe where I still feel and hear with crystal clarity the kick and metallic action of an R4 fired in anger. In which the war was a motorcycle race or a rugby game, and the adrenaline rush is a stronger memory than the faces of dead friends or the smell of dead swaps.

I had my first glimmer of doubt about who the real enemy was long before that evening in Mahanene. We were loading a young dead swap onto a vehicle after a contact one day. Someone's hand slid along his side and the black pigment was stripped back, exactly like paint does if you push your finger along the surface before it dries. The layer of skin underneath was white as snow. It surprised the shit out of me. But it took another twenty-odd years of living in apartheid era South Africa, Kenya, the USA, and the former USSR to understand the symbolism of that moment and what it meant.

I have learned that the real enemy is not the oke with the funny accent, the different skin colour, or the strange customs. The real enemy is the politicians of all creeds and colours who play human chess in their insatiable lust for power. For MONEY, when you filter the bullshit and stirring speeches.

The British band Black Sabbath was one of my favourites during my high school years in the late 1970s. Neither the ominous titles to some of their songs, nor the lyrics of *War Pigs* held any significance for me then.

'...Politicians hide themselves away
They only started the war.
Why should they go out to fight?
They leave that role to the poor, yeah.

Time will tell on their power minds,
making war just for fun.
Treating people just like pawns in chess,
wait till their judgement day comes, yeah...'

Now I understand. Politicians. Fucking war pigs. Fuck 'em all.

53
Pilatos Dos Santos

Throughout history, short acronyms have always had a disproportionate impact on peoples' lives. Ask any young man roaming the streets of 18th century London what RN meant. Or any 1930s European Jew what SS meant. In the SADF, our lives were tangled up in TB, PB, R1, R4, JL, HE, AT, AP, WP and PT. And, of course, PF.

Pee-fucking-Eff. I doubt any other two letters in the alphabet evoked as much emotion. Strictly speaking, it meant 'Permanent Force' or professional soldiers, but in reality was mostly used as a derogatory term for disrespected or disliked individuals. I remember many PFs of various rank with great respect, but also many who were despised by all who served under them.

By definition I was PF myself, having signed up for four years, but I was in the infantry via the SAAF by default and had no use for base camp rondfok. I was interested only in being on ops or civvy street, which made two of the four years torturous to say the least. As a junior NCO, most of my tormentors were Staff-sergeants and Sergeant-majors. Many of these individuals were poorly educated and excessively motivated, some had borderline sadistic personalities. I also encountered numerous officers who were piss-poor leaders, stupid, and blindly ambitious. In Ovamboland I used up at least two of my nine lives because of one such fool.

Infantry School had many highly regarded professional soldiers on staff. My company CO was Captain Peter Rose, a gnomish oke whose short stature belied his outstanding leadership skills. He came to us from 32 Bn and replaced a sadistic, arrogant officer named Lotheringen who once shot at an instructor named Corporal Moore while he was setting up targets for a weapons demonstration. I always suspected he got the chop for that stupid stunt. That, and the fact that Alpha was the kakkest company in the unit because of his poor leadership. The only thing I remember him ever saying to us was '... *julle sleg bliksems*' over and again, like a stuck record.

Captain Rose, in contrast, gathered the company and said: '... *manne, what we have here is a somfu. We must fix it, otherwise we'll all be pilatos dos santos.*' Captain Rose was a real soldier and leader, and we soon understood his English-Afrikaans-Portuguese slang very well. Small wonder then, that he turned it all around in spectacular fashion with help from Asterix, the CSM. Asterix stands out in my memory as an exceptional NCO, respected and well-liked. Those two transformed Alpha Company from a *self-organising military fuck up* to the top company at Infantry School in 1980. Needless to say, much of the effort was ours. They ran us till we were *piel in die sand* just about every day.

There were men of this calibre in other companies too. I specifically remember a Sergeant-major Steenkamp, who had earned the Honoris Crux at Bridge 14 in Angola in 1975. He was an intimidating figure: very tall, olive-skinned with a long face, dark eyes and a black moustache. I didn't have any first-hand encounters with him, but he always struck me as a bit aloof. His body language said that the noise and posturing many NCOs used as intimidation tools amused him, but he couldn't say so out loud. Some of the unit's HQ officers were distinguished soldiers too. Commandants Serfontein, Thirion and Holtzhausen were people we saw only at battalion parades but were talked about with much respect. No doubt there were some idiots too, but I was lucky enough not to have dealings with them while in Oudtshoorn.

The first PF whose acquaintance I made at Berede was a comical caricature of a Sergeant. He was blonde, short and stocky with no neck, a sloping forehead and a fu-man-chu moustache. He was also an arse-kissing loudmouth and dumb as a bag of spanners. Us new Corporals had lots of laughs baiting him once we figured him out. The rest of the rank at the unit was a mix of respected, capable individuals and despised, incompetent fools.

I firmly believe that anything worth doing is worth doing to excess. That's why I didn't mess around with lower ranks but *sommer* made enemies of the 2IC, a Major we called 'Ratsy' or 'Bugsy', and the RSM.

Major Ratsy was a nasty little man, short and chubby, with close-set piggy eyes and a black Hitler moustache. His brown pants stretched around his tree-trunk thighs and crawled up his arse like some obscene knife through a slab of cheese. Bike squad had too many Corporal instructors at one point, so he made me the unit's Signals NCO. I hated every minute of it. My attitude was, the only reason to be in an army is to fight a war, so I applied for a transfer to 32 Bn. When he found out, Ratsy personally yanked my file and tore up the papers. My uncooperative attitude eventually made him look bad to the CO and he chased me back to bike squad. A few years later he got demoted for organising an assault on another unit's officers at the Officers Mess in Potchefstroom military base. A real gem. I fucking despised him.

The RSM was a typical base camp marionette with a handlebar moustache and chains in the bottom of his pants legs to make them hang straight. When he did an *omkeer* or halted, it sounded like someone dropped a socket set. He looked like he had a *Brasso'd* parade stick surgically attached to his left armpit. To the best of my knowledge he never spent one day on patrol or one night in a TB. Funny thing is, I dodged this oke for two years but for the life of me I can't remember his real name. We Corporals called him 'Fuck-knuckle'. We neither feared nor respected him and regarded him merely as a nuisance. Not once in those two years did I have a normal conversation with Fuck-knuckle who, in reality, was my boss two levels up the food chain. He was a sad reflection of how dysfunctional some of the SADF leadership was.

Until spring 1983, the CO of bike squad was Captain Jon Stroebel. He was blonde with pale skin that turned red in the sun, of average height with Popeye forearms, and an excitable and volatile personality. Jon won the 200cc class in the Roof of Africa a time or two so he had lots of credibility around bike squad. Despite his daily tantrums we respected and liked him. He left the army and went to work for BMW's motorsports division. I am told that he died several years ago. Jon Stroebel was a good oke.

The CO of Berede was a dark-haired Commandant with a hangdog expression, a very decent man whom I unfortunately crossed swords with during my last few days in the SADF. By that time I was so *gatvol* and focused on getting out that I didn't appreciate the CO's good qualities, and ended up pissing him off. I regret that, but it's still a funny story.

The unit's paymaster was a lanky admin Captain whom we juniors and short-timers ridiculed and disliked. When I resigned, some clerk somewhere misread the date and I got my final paycheck a month early. I was reasonable about it and politely asked the Staff-sergeant in the pay office to please sort it out. Nothing happened. After getting the same fucking excuse for three weeks, my patience was worn thin. I was broke, and both the fat pencil-neck Staff and his skinny cunt boss seemed to enjoy my predicament.

The morning of my last Friday in the army, Patrick the bike squad 2IC and I took two demo bikes out to the motocross track. Me out of frustration, him because he was my friend. Or, more likely, just wanted to go fuck around too. The demo bikes were brand-new XR 200s donated to the Demo Team by Honda. They used the little XRs to perform very impressive displays of riding skill at venues like the Rand Show and the Durban Military Tattoo. The demo bikes were strictly off-limits and jealously guarded by the fat, non-riding PF loot who was bike squad CO at the time. I remembered him as Sergeant Black, the Alpha Company quartermaster at Oudtshoorn in 1980.

Staff-sergeant Tourney, a hard-arse SWASPES tracker on the bike we allowed him to ride, and the reason he was semi-famous.

He was a cunt in 1980 and still was, three years later. However, I had less than a week left and was one signature short of completely klaared out. I felt fuckall.

Upon our return to camp this wanker went through some noisy gyrations, then threw a huge stack of files at me. It was the maintenance records of the pile of wrecked bikes that had been used up in training over the previous few years. Out of pure malice he gave me some bullshit admin assignment that would keep me in the workshop for my last weekend in the SADF. I'd finally had enough and flat-out refused. I told him '...*You lot don't pay me, so I don't work.*' To my surprise that didn't go over well. After he finished jumping up and down, he asked if I would say the same to the CO.

What a stupid fucking question.

That's how I ended up on the CO's carpet after lunch, with him yelling that he'd chuck me in the DB and I wouldn't klaar out in a week. For many weeks, in fact. All I said was '...*yes sir no sir*' and thinking '...*three bags full sir*' while he ranted at me. After he ran out of breath and threats, I made an *omme-keer* and left. An hour later a clerk walked into the bike office and told me that my money was in the bank.

The same cast of characters performed in a bit of drama we'd had earlier when one of the demo bikes disappeared. I had become close friends with Patrick, a loot who joined the unit while I was in Ovamboland. He had been a rifleman in platoon 6 in 1980-81 before joining the PF and going through Infantry School. We both had a lot of tracking experience. We easily found the spoor where two troops had pushed the bike out of the workshop, down a dirt road to the fence along the main road about a kilometre away. One person on the outside helped lift the bike over the fence, loaded it onto a bakkie, and drove off.

We could almost tell what colour eyes the thieves had from the crystal-clear, undisturbed spoor. We tracked them back to the bungalows, where the army's obsession with sweeping dirt prevented us from pinpointing which building they returned to. When we reported our findings to ex-Sergeant Black and assorted PFs gathered at the bike office, they said we were talking kak so we laughed at them and left them to chase their tails. That incident stands out in my memory as a prime example of the corrosive effects the stupidity and arrogance of some PFs had on the overall operation.

In hindsight, it seems that most of the PFs worthy of respect at Berede were horse people not very concerned with the military side of things. Like the German CO of the riding wing, Commandant Peter Stark. He was rumoured to be an old poacher from SWA and was an absolute no-nonsense type. He was generally friendly and seemed amused by us bike hooligans, but we heard many stories about his short temper and perfectionist approach to riding and training horses. He supposedly punched a stubborn horse one day and knocked it out cold. The Commandant had two sons my age in the army, one was a berede platoon commander and our drinking partner. He also had a shapely legal-age daughter whom we saw riding or walking around in skin-tight riding breeches most days.

The unit had several NCO farriers and other skilled tradesmen who didn't have a saak with the army, all they cared about was horses. One old Sergeant-major whom we called 'Crinkle-Cut' was utterly out of place in the military and left the stables only at tea time, lunch time, and knocking-off time. He had his routine perfected and drove out the gate at five o'clock sharp every day. Old Crinkle-Cut was a farrier of great repute, deeply wrinkled from years of shoeing horses in all kinds of weather. He looked like a hairless Shar-Pei but he was a friendly, if not very talkative, old goat. We respected him for his dedication to and expertise in his trade.

In contrast to Berede, I don't recall any real fuckups at SWASPES. We never saw the CO, the RSM was fairly easygoing, and the two

Captains in charge of Bike Squad and Ops Wing respectively were decent okes too. The only PF we knocked heads with was a tracker Staff-sergeant named Tourney. I don't remember what his function was, only that he could be difficult at times. He was slightly notorious for having been on the cover of *Soldier of Fortune*, a gun-crazy American magazine that was flourishing with all the wars going on around the world at the time. Despite the occasional clash, we respected Staff Tourney. I remember chatting quite amiably in the pub with him about SAAF pilot selection, which he considered doing but at 25 was too old for. He annexed an XR 500 for himself and seemed to ease up on us from then on, presumably out of shared experience.

We befriended a one-armed ex-Rhodesian ex-Grey Scout Sergeant-major at Spes and spent several fun drunken evenings with him. Another good oke.

There were a few more PF tracker NCOs at Spes, hard men who were very good at their job. Sergeant Stony was a huge bear of a man with woolly red hair, very aggressive but a nice oke off duty. Our drinking buddy Dave was a short and lean blonde Sergeant, whose girlfriend was a cousin of one of my high school friends whom I later ran into at Ombalantu. Some months after returning to SA I sat in my parents' living room and watched the SABC TV newsman announce in somber tones that Staff-sergeant Dave Ward had been killed in action.

Paradoxically, we clashed with most of the PFs we came in contact with in Ovamboland. One would expect more professionalism and less *rondfok* the closer to the sharp end you got, but it wasn't quite so. The exceptions were the CO of 52 Bn, Commandant Koen whom we only saw once or twice, and the RSM, Asterix. The average PF mentality saw us bikers as real scum and treated us accordingly. We were *specialist* if not *special* forces, and ordinary infantry officers didn't know how to handle us or use us effectively. I always wished we could have been attached to Koevoet in Ovamboland, but that was too revolutionary for the rigid thinkers who were in charge of the infantry battalions.

At every base we made at least one enemy among the jam stealers. At Ogongo it was the battalion 2IC, an Intelligence Corps Captain who was a real menace and almost got me and others killed several times with his stupidity. At both Okalongo and Ombalantu we got cross-wise with Majors from SAI units and were thrown out of 'their' bases. Cunts.

At Mahanene we fell under a good man, the 3 SAI company commander Captain Rocha, but the CSM and one of the chaplains were bad news.

With 25 years of hindsight, I clearly see that there were two types of people in the army: the real soldiers, who were good at their jobs and whom we trusted and respected, and the 'PFs'. Those individuals who ducked, dived and connived safely in base camp, who lived for rondfok and seemed oblivious to the fact that there was a war going on around them.

Despite all the training and conditioning, I never really hated SWAPO. They were the enemy so our job was to kill them, pure and simple. We actually respected them to a certain extent. Those who were supposedly on our side but spent their time fucking us around and needlessly put us in harm's way with their stupidity, now *that's* another story.

I fucking despised them for betraying our trust. Still do.

54
Pukka Gen

The official name of the army unit in which I spent three intense and formative years of my life, was *SADF Equestrian Centre Motorcycle Wing*. Not hard to see why everybody except most PFs called it *Bike Squad*.

It was a small part of Equestrian Centre, based on a farm 13 kilometres north of Potchefstroom on the Ventersdorp road. The unit's primary focus was training mounted infantry and their horses for the bush war. Operational Wing, the official name for the horse platoons, was much bigger than bike squad and usually had four or more platoons in training. A separate Equestrian Wing consisted of expert horsemen and –women who participated in dressage and show jumping competitions at the highest level.

The first bike platoons were assembled in 1977 when SWASPES was formed. SWASPES, short for 'South West African Specialist Unit' was the battalion-sized SWA Territorial Force (SWATF) unit in the operational area from where bikes, horses, trackers, and dogs were deployed. The horse and bike platoons were numbered sequentially. My time in bike squad spanned platoons 10 to 13. Platoon 10 was the July 1980 intake that left for Ovamboland three months after my arrival. In early 1981 I was involved in training platoons 11 and 12 for a short time. I would eventually spend a year in the bush with them, and my memories of bike squad revolve around those okes. Platoon 13 went to Ovamboland just before I left the SADF. For a while I was sorely tempted to go with them, and with 25 years of hindsight I sometimes wish I had. But I had been disappointed and disillusioned once too often, and my desire to be free and get on with life was stronger.

Bike squad troops were selected from volunteers at the SAI units at the same time as recruits for specialist units like Infantry School, parabats and recces. The unit attracted motorcyclists of all kinds, especially dirtbikers, and had many excellent riders in the operational platoons. However, some had no motorcycle experience and learned to ride during the excellent six-week crash course they went through at Potch. Pun intended, I was one of them. My instructors were national level motocross racers, and I have no doubt that those painful weeks saved me years of learning and crashing on my own.

Motorcycle training at Equestrian Centre, Potchefstroom.
Standing at left without helmet is the CO Captain Jon Stroebel.

Many top national level riders from various disciplines, especially motocross and offroad racing, ended up in bike squad. The CO for most of the time I was there, Captain Jon Stroebel, actively recruited them from whichever units they were called up to and had them transferred to bike squad. They were employed in Potch as riding instructors and Demo Team, while the platoons went to Ovamboland. The Demo Team travelled around the country giving amazing displays of riding skill at venues like the Rand Show, the Durban Military Tattoo, and agriculture shows. It also allowed them to compete on weekends and fulfill their sponsorship obligations.

Some very fast and skilled riders did go to the platoons. The two best riders in platoon 12 were a section leader and a rifleman who raced motocross at national level.

During my three years in bike squad we rode mostly Hondas. Until 1982, training at Potch was done on a fleet of much-abused Honda XL 350s and Suzuki SP 370s. Both these bikes were heavy and underpowered and not suited to the harsh terrain of Africa. Only XL 350s and various XR models were used operationally. We rode new but obsolete 'bigwheel' Honda XR 500s for the first part of 1982. While tough enough for the conditions and terrain, they were not modern dirtbikes and handled poorly. All that changed with the batch of brand new XR 500R 'Prolink' single-shock Hondas we got in May 1982. They were excellent bikes, handled well, and Honda's typical reliability made them ideally suited for the job.

Operational bike troops went through standard infantry training before doing the riding course and motorcycle operational training. On paper the platoons consisted of a platoon commander, a platoon sergeant, and three sections of ten men each including section leaders. In Ovamboland and Angola we operated as two equal-strength sections of 12 to 15 men each, as the platoon was never at full strength for a variety of reasons. We evolved a system of a section leader in direct control of each section, and the loot and me in overall operational control. It only took a few hand signals to split into two autonomous small units commanded by the loot and me respectively, when required. It worked very well and was typical of the improvisation that made us more efficient in the bush.

Another was the way we carried our weapons, which evolved from my feeling very vulnerable in imminent contact one day during Ops Yahoo when I realised it would take several seconds of frantic gymnastics to get my R4 into action. From that day I slung it barrel down to my right so I could simply swing it under my arm and start shooting chop-chop. Most of the platoon followed suit. Years later I found out that earlier platoons had done the same thing for the same reason, but the knowledge wasn't transferred or incorporated into motorcycle doctrine.

Attrition in the ops platoons was high. From start of training to end of operational service it ran as high as 75-90%. A large percentage didn't make it through initial motorcycle training and individuals were weeded out by injury, personal problems at home, and other reasons as time went by. Many just disappeared from the platoons for unknown reasons, presumably by request.

Motorcycles and Ovamboland were made for each other. The terrain is mostly open, absolutely flat, and has no rocks from what we were told and observed. It is ideal motorcycle terrain. The only thing more mobile than bikes in Ovambo were helicopters. Buffels, Casspirs, and even Ratels couldn't keep our dust in sight for long. In fairly open terrain we could average 90 km/h and did as much as 120 when needed or we felt like it. However, as illustrated in the area around Otavi and Tsumeb, bikes were not effective in dense bush.

Bigwheel Honda XR 500s on a farm tweespoor west of Otavi during Ops Yahoo. This shot clearly illustrates the limitations of motorcycles in dense bush.

Bike platoons were, in ancient Greek or Roman terms, 'light' infantry. The incredible mobility of modern dirtbikes made the platoons an ideal quick reaction force, and they were mostly used in that capacity. We carried no LMGs as in a standard infantry platoon. Each man had his 5,56 mm R4 with six or more magazines.

The standard magazine held 35 rounds, but we liked having one of the bigger 50 round mags on our rifles and the standard ones in our webbings. Experience soon taught us to only load 45 rounds, with two or three tracers near the bottom to alert you that the magazine was almost empty. More than 45 rounds compressed the spring too much and the R4 would actually jam, as shocking a concept as that was.

One rifleman in each section carried a *snotneus*, the M79 40mm grenade launcher. After an informal Sunday morning shootout in the *asgat* at Ogongo, three okes who were deadly accurate with the funny looking little weapon were selected to carry them. After a while they left their R4s behind and carried only the *snotneus* and enough ammo to wipe out a battalion. Most of us carried hand grenades of some kind: M26 HE grenades, various coloured smoke grenades, and a few white phosphorous grenades. I vaguely recall carrying a Claymore mine once or twice, but it was too heavy and not suited to our operations so it was left behind. Two dozen bikes roaring up to some spot in the bush to lay an ambush was not a viable tactic.

The loot and all three section leaders initially carried A53 VHF radios but that also went down to two after a while, one in each section. The single most hated piece of equipment was the big and heavy B25 HF radio. The loot always picked someone to carry it. As I recall, whomever had caused shit of some kind usually got stuck with it.

We travelled fast and light. Each man carried only his sleeping bag, a groundsheet, a few tins of food, two or three water bottles, and ammo. The only personal things I had space for was my little 110 camera in my webbing, a Swiss Army knife on a string in the front pocket of my brown pants, and my bush hat. Most okes carried a *naai-fuck-n-spoon set* or *pikstel*, the army-issue three-into-one cutlery set, but I used my pocket knife to open tins and eat whatever was in them. Labels tended to get rubbed off and all the tins looked the same. Meals often reenacted Oliver Twist's birthday.

The only toiletry we carried was 'white gold', aka bog roll, shit paper, *kakpapier*. Washing, or any kind of hygiene, was officially prohibited because SWAPO supposedly could smell soap or toothpaste in the bush. We simply got dirtier as the days went by. On one month-long operation in Angola, I got so filthy that I made notes on my dirt and grease-blackened pants legs with twigs during order groups. I threw away my socks and underpants after about two weeks, when I schemed SWAPO could smell them twice as far as any aftershave in existence. I realised how filthy I was when the pong stayed with me at 110 on the tar road on the ride back to Ogongo. It took many hours of scrubbing to get our bodies and clothes clean after such capers.

Vest-type webbings were standard issue for all SWASPES units, but a few okes preferred chest webbings they scrounged from somewhere or made themselves. We had all been issued with red-and-black motocross boots at Berede, but seldom wore them. They were good for training and riding but totally unsuitable for operations, where you did lots of walking, bike squad or not.

Grensvegter! On one of our first patrols,
in the charming village of Marco Quinze in Angola.

We were issued with old-style pisspot helmets at Spes and wore them religiously. On our first patrol or two in Ovamboland we wore bush hats, but had a few mishaps and went back to wearing helmets. The standard issue Spitfire pilot style goggles were not popular. Besides the uncool look they were too restrictive. Most okes brought, or had sent, motocross goggles from home. I had a pair of yellow and red SCOTTS that served me well all year. We also discovered early on that gloves were essential. Since we were not issued any, everyone brought their own and we soon identified individuals by their gloves. My father sent me a pair of dandy white leather gloves which I disliked in the beginning but learned to appreciate greatly as time went by.

We only did basic maintenance in the field. We cleaned the air filters and changed many sprockets and chains ourselves but oil changes, timing chain adjustments, and specialised maintenance was done by tiffies. The XR 500's Achilles heel was the timing chain, which stretched from hard use and made a very distinctive slapping noise against the side of the cylinder. We had three spare bikes at Ogongo and eventually used them, after a few bikes threw timing chains or developed other problems. The Hondas tolerated about three months of hard use without much maintenance before developing mechanical problems.

Of course, everybody made small modifications or adjustments to their bikes. Some, like removing the speedos, were unplanned and happened when wheelies went bad. We experimented with the exhaust baffles, removing them made the bikes much more responsive but significantly louder. In Ovamboland we did some informal tests one day and found that the benefits of higher performance justified the increased noise level, so we all took the baffles out. A single bike could be heard at about 100 metres depending on the wind. A section or more would cause a rumble at 300 metres, but the direction was hard to determine. We also found that disconnecting the headlights made a huge difference in how far we could be detected. With lights on, bikes were visible at two or more kilometres in that flat landscape. Without the lights the brown bikes, especially if coming towards you, were hard to detect until about the time you heard them. The lights were wired to be on when the engine ran, so we disconnected all headlights permanently. These measures were reported as very effective by a captured swap at Mahanene base.

Our biggest limitation was petrol. The XR 500's petrol tank held nine litres and it could do about 180 kays on a tank. That equated to two days of easy, or one of hard riding. The only times we had fuel available outside the bases were during the week we spent with the Noddy cars, and on one operation in Angola with other units when we stayed out for a month. We never ran out of petrol, after months of riding the XR 500 we were able to judge distance and consumption very well. Strangely, I can't recall a single instance of fuelling the bikes, even though we did it roughly every second day.

During training, standard infantry drills were practiced to distraction. We had to do a COIN refresher course at Oshivelo before they would allow us into Ovamboland. Running around like infantry twats, laying ambushes and setting up TBs, was a waste of our time without bikes. Ops Yahoo had just ended, during which platoon 11 gained much operational follow-up and tracking experience. Platoon 12's original leadership were both July intake and they klaared out of Spes in June. A new loot and platoon sergeant arrived from SA and I was moved from 11 to 12. I only realised many years later that I was assigned to platoon 12 so both platoons would have one operationally experienced oke in charge.

The only motorcycle drill I remember practicing was the ambush drill. It was a sporty affair on bikes. We'd cruise around in formation until someone, usually some PF 'evaluating' us, threw a thunderflash or emptied a magazine over our heads. When the 'ambush' was sprung, the section that drew fire would drop their bikes and revert to the basic infantry action of *dash-down-crawl-observe-sights-fire*. Meanwhile the other section jaaged around one side to flank the ambush, and attacked with good old fire-and-movement. The goal was to establish an L on two sides of the enemy, and we based our formations and tactics on that principle. Obviously, this drill was useless in the dense bush around SWASPES, but there we gained other valuable experience.

Ambush drills accounted for lots of lost blood and skin. In the heat of the moment you tended to forget what speed you were doing, bail out when shots were fired, and hit the ground in a ball of dust. When we arrived at Ogongo Asterix, the 52 Bn RSM, wanted to see us in action so we did a live ammo ambush drill for his benefit. It went spectacularly well, judging by the amount of dust we created and the amount of blood and ammo we expended. Asterix was duly impressed.

As it turned out, landmines were a much bigger threat than ambushes. My platoon didn't get ambushed even once, as far as we knew. All that bailing off moving bikes and subsequent bleeding was for naught. We usually travelled at such high speeds in the open terrain of 52 Bn's area that it kept us out of trouble.

Mines were a different story. Both platoons, 12 in 52 Bn at Ogongo, and 11 in 53 Bn at Oshigambo, had their share of extremely close calls but were lucky enough to come away unscathed. Had we stayed longer, the odds would eventually have caught up with us. A landmine is a terrifying weapon because of its insidious nature, and especially so in Ovamboland. Literally every square metre of that sandy region potentially hid a mine, and eliminating the threat was impossible. We did what we could to minimise the risk, and lived by the principle of '*jou beurt is jou beurt*'. Avoiding roads reduced the odds of tangling with anti-tank mines, but the stupidity of some PFs exposed us to needless danger.

Line abreast section of a T-formation east of Okalongo, October 1982.
6 foot 4 Dave closest to the camera and Fourie beyond him.

Our biggest worry was the Russian TM 46 and TM 57 'cheese' mines that contained six kg of explosives each. One was enough to blow up a tank, and SWAPO had the nasty habit of planting two in one hole in an attempt to destroy the very effective mine-proof vehicles the SADF used. There is no such thing as a mine-proof bike, as we were all too aware after bike squad's only operational death in January 1981.

Bike squad was one of the unique aspects of the bush war. Motorcycles have been used extensively in communications and transport roles since the First World War. Despite extensive research, I am not aware of any other motorcycle units that deliberately rode into contact like motorised cavalry. The closest incarnation occurred in the German army in the Second World War, but they did not ride into contact like we did. Bike squad in the form I knew it, only existed for about eight years and the 500-odd men who served in it during that time are largely forgotten. Even with the tremendous resources of the internet I have been unable to find any information about bike squad, except for a few vague references in individual accounts of the war and SADF organisational charts.

That's not right. I will fix that by researching and writing a detailed unit history. There are many more stories like mine out there that must be recorded, if for no other reason than scoring free beers in the pub.

55
Fog Lifting

Supposedly it takes 50 years for historians to separate fact from fiction from propaganda, and get an accurate picture of what really happened at some specific time. Halfway there, more and more information about the bush war is coming to light but it still only scratches the surface. Not to mention, my generation's war is now *subject non grata*. People just don't want to hear about it. It makes them uncomfortable, for many former government fucks hopefully because of a guilty conscience. I'm not holding my breath.

Having lived as an ex-pat since 1984, I only started sorting out the events of a generation ago in 2006. I've got some information off the internet, but much of that is rubbish. Opinions are like arseholes, everybody has one. Credibly researched histories and first-hand accounts are still few and far between. I'm not aware of more than fifteen or so, but those I've seen have all added bits and pieces to the puzzle.

An author named Peter Stiff has written a trilogy of books full of facts, dates and names that so far enlightened me the most. Reading *Covert War*, his history of Koevoet, was a real eye-opener. I spent many weeks operating with them in 1980 and 1982 and my experiences match his narrative exactly. I read all about the incursion south of Ovamboland by SWAPO's special *Volcano* unit in 1982, as well as Ops Yahoo, the SADF campaign to wipe them out. At the time we didn't know much about the big picture, or most of what went on outside our private little war. I learned much from Mr. Stiff's book.

The overall commander was then-Commandant Serfontein, whom I had seen at Infantry School two years before but never had any dealings with. The incursion started around 25 March and ended in early June, by which time all the terrs had either been killed or captured. Between 245 and 294 had infiltrated, of which three groups of around 40 each reached the south. It was the largest incursion of the war. In addition to a parabat Captain who was killed by a grenade near Otavi, one farmer died in a landmine incident, and another farmer and the soldier guarding his farm where killed by terrs heading back to Angola. They killed at least one farm worker too. On the first day of Ops Yahoo a Ratel was shot out with RPG-7s near Tsintsabis and eight men were killed. A Ratel crewman was killed when the vehicle rolled just south of Oshivelo, and another soldier died when he stepped on a boosted anti-personnel mine on the cutline 40 km west of Tsintsabis. The same cutline we rode numerous times, and the camper Colonel tried to make us sweep every day until we taught him a hard lesson about soft sand.

The loot and Trieloff discussing the lack of mud in Angola above beacon 7.

Accounts of contacts, mine incidents and major battles all over Ovamboland and Angola brought back many memories. Sam Nujoma's home of Ogandjera, Tsandi, Ombalantu, Mahanene,

Ongenga, Okalongo, Oshikuku, Kwambi, Marco Quinze, Dombondola dam and Ruacana are all images in my head. Beacons 5, 7, 9, 10, 12, 15 are mopani-bushed reference points on the cutline that look identical in my photos.

But I also learned of the betrayal of 3000 Ovambo Koevoet policemen by the SA government, specifically FW de Klerk and his lackeys. They did the same to 32 Battalion, which was largely made up of former Angolan FNLA fighters. And 31 Battalion, the much-abused Bushmen. Each and every one of us who fought their war for them, when you think about it.

Several levels removed from my on-the-ground perspective, the strategic view if you will, is the book *Days of the Generals* by Hilton Hamman. An intense read, it takes a lot of concentration to digest. While I disagree with some of the generals' opinions of Koevoet specifically, it clearly illustrates the hypocrisy, stupidity and giant egos of those in power at the time. More importantly, I learned that the people in charge of the SADF were professional soldiers and not politicians. It seems that the generals had honourable intentions, which brings me some peace of mind.

The negative experiences I had in the bush occurred primarily at company and battalion level, as demonstrated by the captains and majors we clashed with in Ovamboland in 1982. I saw much stupidity at that level. In the years since I have observed many times that 'middle management' is the weak link in any big organisation. Even the best ideas and plans are worthless if executed poorly by those in charge of operations on the ground.

Twenty-odd years later, statistics about the bush war are also widely available. I learned that 1982 was the SADF's bloodiest year of the war with some 228 soldiers, airmen and sailors killed. 194 died in 1980, when I had my first taste of the bush war with Infantry School and Koevoet. 1983 and 1987 each killed 205. In total, some 2400 men died in the war against the *Rooi Gevaar* between 1966 and 1989, the official beginning and end of the bush war.

My first SADF memory is of klaaring in on 4 February 1980. As with any job, I had to fill in and sign a mountain of paperwork. The part of that first day I remember very clearly was making out my will. A *pencil tiffy* Lance-Corporal sporting a *kapoen* beret, comfortably seated behind a green aluminium folding table next to a rugby field in the late afternoon sun, was nagging me to finish. I was 17 and didn't have a pot to piss in or anybody to leave it to. After some deliberation I bequeathed the few thousand Rand paid on the occasion of my death in action or accident, based on my monthly salary of R267, to my father. I figured he would find a use for it. The Lance-Corporal said it was a good choice.

My last minutes in the SADF are as crystal clear. My friend Ikes who had helped me terrorise Ballas, the *dominee's* beagle at SWASPES, and I drove out the gate at Berede together on the first day of February 1984. We had both obtained the last signature on our klaaring-out forms and dodged the RSM by sneaking past the HQ, behind the shadow-net parking stalls that lined the 200 metres of road to the gate. I stopped just outside the boom and we stripped off our browns and changed into shorts, T-shirts and flip-flops. A Joe Walsh tape was cranked up full volume in my bakkie and *Life's Been Good* echoed across the veldt. Our first act as civilians was to pull down our shorts and gooi brown eyes back up the entrance road, at the end of which was the RSM's office window. A Samil 50 full of troops drove through the gate at that moment and they laughed and cheered with great enthusiasm.

Despite the hilarity and excitement of the moment, I felt strongly apprehensive. I remember thinking '...*now what?*'. Ever since that day I wrote my will six weeks after leaving high school, I had been conditioned, and assumed, that I would probably not survive to be a civvy with years of opportunity ahead of me. But the thought of those same years stretching out endlessly safe and dull caused me even more distress. I was 21 and suspected that my life had been changed permanently. I stood bent over outside that boom gate with a warm breeze tickling my white arse and wondered what was to become of me.

It turned out all right. Life has come full circle in a strange and circuitous way, and I have done well. I will never go to war again for or against any country or ideology. The only cause I will ever take up arms for again, is my family and friends. I now know what's worth fighting and dying for, and a flag of any colour it's not. That's all.

56
I Wanna Be A Bike Squad Rider

A song we learned from the ex-Rhodesian Grey Scout members of SWASPES was adapted for use by the berede and bike platoons. We never sat around campfires singing songs like boy scouts, but I still remember the words:

'*...I wanna be a bike squad rider*
  live my life of sex and danger
  I wanna go to north Angola
  kill that fucker Sam Nujoma...'

In 2001 my old bike squad friend JB and I went on the Bike SA Desert Run, an organised ride through the Northern Cape and Namibia. After it ended in Swakopmund, we toured some of the sights of that beautiful country. We spent a cold night camping in Luderitz, an old German mining and fishing town on the southern Atlantic coast.

The next morning we went to the larniest hotel in the little town for tea and crumpets before heading back to Johannesburg. It was a nice day with sunny, clear skies and a cold breeze blowing off the icy ocean. We sat at a table on the verandah, separated from the water's edge by a perfect green lawn. Guards at the gate had told us that a government delegation was on its way to the hotel for a conference of some kind.

I was on my second cup when four soldiers armed with AK-47s came walking around the corner and casually took up positions on the lawn, clearly enjoying the scenery while standing guard. Soon after we heard voices and movement on the balcony behind us. The government delegation had arrived.

We both turned around and looked up. Several well-dressed men stood on the balcony five metres away. I thought I recognised one bearded face. He stood at the railing looking down at us and I made eye contact with Sam Nujoma, the President of Namibia. We stared at each other for long seconds as he subtracted twenty years from

our ages and recognised us as old soldiers and enemies. I grinned at
him and he turned and sat down.

I wanna be a bike squad rider faintly sounded, I thought, and I
looked across at JB. It wasn't him singing. He caught my eye and we
just smiled.

Basic riding course, Potchefstroom 1982.

Grant Petersen, the Platoon 12 platoon sergeant whom I replaced when he klaared out in July 1982. He won a beer by chasing this goat halfway to the mountain in the background.

Training ride with platoon 12 and corporal Petersen at right heading back to base with a flat front tyre. Foreground facing the camera is JT du Plessis, at right is Banks' backside and Steyn the Wheelie King.

Training on the new XR 500s, still with intact paint, at SWASPES. Rifleman Banks closest to the camera was 16 years old but his style was better than that of the 19-year-old loot behind.

Training in the gravel quarry just east of SWASPES base.

Training in the gravel quarry just east of SWASPES base.

Halfway between Otavi and Grootfontein was a mine named Kombat. Just north of the mine was a sandy motocross track that became our sand training facility. Here the platoon is leaving the track, headed back to SWASPES.

Sharpening my sand riding skills at Kombat MX track.

171

Races at Tsintsabis. I learned about tyre pressure as it relates to soft sand on this occasion.

On the Bravo cutline between Tsintsabis and Oshivelo.

On the tar road between Tsumeb and Otavi, returning to SWASPES after we gave a Skiet Piet Colonel at Tsintsabis some sand riding lessons.

2/Lt Hagen and me at Oshivelo before a live-ammo night ambush 'evaluation' after ten wasted days of COIN training without our bikes.

Returning to Swaspes from Oshivelo by Kwêvoël.

Returning to Swaspes from Oshivelo by Kwêvoël.

Fixing flats in Ovamboland. Note the typical kraal and large Ford bakkie popular with the PBs. We borrowed the red foot pump from the family in this kraal.

We went for a swim whenever we found dams in Ovamboland and Angola. Half the platoon swam while the other half stood guard.

*We went for a swim whenever we found dams in Ovamboland and Angola.
Half the platoon swam while the other half stood guard.*

*Our midday siestas got boring so I started chasing donkeys and pigs for
amusement. Le Roux on the donkey.*

Where the main road wasn't tarred, we rode alongside to avoid landmines. The loot and section leader Trieloff stuck in what they thought was a dry riverbed at Oshikuku, east of Ogongo.

Spread out formation on a wide open shona south of Oshakati, Ingalls in front.

One of the typical hollows we liked to TB in, north-east of Mahanene.

Operating with the Noddys.

A chance meeting with berede platoon 28 on the Dombondola shona.

Siesta in a hollow east of Mahanene.

40 Days party at Ogongo. L to R: Maree, JT du Plessis, Ingalls.

Braaiing a goat we bought from a PB in Angola above beacons 9 and 10. At rear, L to R: 6 foot 4 Dave, Hagen, Vorster, Fourie. Front, L to R: Banks, De Jager, Ingalls, Maree.

Mounting up after a relaxing braai in Angola. Corporal Fourie in left foreground, Steyn the Wheelie King far left.

Bruce McKellar from Durban, Bike Squad Demo Team 1983/4. RIP

Demo Team 1984. Bike Squad CO Patrick Devy standing left.

At Durban Deep motocross track near Krugersdorp, Transvaal in 1983 with my Maico 250 and life-long bike squad friend JB. A sleg civvy at the time, he left the SADF six months before me.

Bike testing at the Gerotek vehicle testing facility west of Pretoria, February 1983. Patrick and me at right.

Acceleration and sound tests at Gerotek.

Part of the concrete loop at Gerotek, basically a paved MX track with some radical elevation changes.

Taking a breather on the Gerotek paved track. L to R: Bruce McKellar, the Kriek builder Ron Wood, Sergeant Kippy.

SA Trials champion Mike Deglon demonstrating severe throttle control.

Mike Deglon at Gerotek. Shortly after this photo was taken, he overjumped this downhill and almost landed on the flat. It was the only time I ever saw him rattled on a bike.

185

Incline testing area at Gerotek, steeper from right to left. Easy on bikes, attention-getting on the Samil 20 demonstration ride we were taken for.

The two custom-built Krieks, based on a flattrack design by the top US builder Ron Wood. A 600cc Rotax engine and 19-inch wheels front and rear made it a brute, hard to ride even for the very experienced riders in the testing team.

Testing at Riemvasmaak, near Upington. The last photo of the Montesa 360cc two-stroke before it gave up the ghost.

Ron Wood and his two Krieks.

SAND! Me flogging the Yamaha XT 550 at Riemvasmaak. The dry riverbed used as test track eventually killed all the bikes except the two Krieks.

A former platoon 12 troop named Du Toit on a Kriek at Riemvasmaak. The extreme sand conditions gave even the SA Trials and MX champions in the group trouble.

Deglon finally murders the Suzuki DR 500.

Ride till you die. On me biffday 39 years later, shortly after a minor crash near my home in the Arizona desert.

Glossary

A

A53 -- VHF radio carried by platoon commanders and section leaders

AK -- AK-47

Alpha-Sierra spoor -- distinctive tracks made by SWAPO boots with a chevron pattern sole

asgat -- (Afr) garbage dump outside base, a big hole dug by bulldozer

awol -- absent without leave, a major no-no in any military

B

B25 -- platoon HF radio, big and bulky

babalas -- hangover

ballie -- old man, also see 'toppie'

bakkie -- pickup truck (Afr)

ballas -- balls, testicles (Afr slang)

ballas bak -- lit baking balls, relaxing, lying around

balsak -- lit. ball bag (Afr), large kit bag issued to all troops

bats -- parabats, ie. parachute battalion

Berede -- SADF Equestrian Centre, or generic name for horse mounted infantry (Afr)

beurt -- turn, chance (Afr)

Bike SA -- Iconic South African motorcycle magazine published since the 1970s

biltong -- traditional South African dried meat

bliksem -- Afr. swear word with many meanings: fall down, hit, punch, beat up etc, same as 'donner' and 'moer'

boep -- paunch, potbelly (Afr)

boere -- lit. farmers (Afr) slang term for SA'ns, especially by SWAPO who considered it a vicious insult

boffin -- expert, good at something, British slang for 'scientist'

bokkop -- lit. buck head, the Springbok head beret badge of the infantry, used as slang for infantry by other army corps

bombshell -- a tactical move, individuals or vehicles scattering 360 degrees like shrapnel from a bursting bomb

bone -- spit-shine, usually boots or shoes

boomslang -- tree snake (Afr) a lethal African snake that lives in trees

boskak -- bush shit (Afr)

bossies -- from 'bosbefok' lit. 'bush-fucked', getting wild and slightly crazy after spending long periods of time in the bush

braai -- (Afr) barbecue, popular South African custom of grilling meat

breeker -- lit. breaker (Afr) slang for tough guy or bully

brown eye -- aka 'moon', bending over and pointing your bare arse at somebody

brunch -- late breakfast/early lunch at 10 am, next meal was dinner at 6 pm

Buffel -- Buffalo (Afr) ubiquitous SADF mine-proof personnel carrier, built on a Mercedes Unimog chassis

bungalow -- general term for accommodation, in this context, barracks

C

camper -- Citizen Force soldier

casevac -- casualty evacuation

Casspir -- mine proof personnel carrier used by Koevoet

cheese mine -- Russian anti-tank landmine filled with 6 kg of explosives

chicken parade -- organised sweep of an area to pick up litter

chicken run -- obstacle on a motocross track consisting of a series of low transverse ridges, requires considerable skill to ride

china -- friend, from Cockney slang 'China plate' = mate

chopper -- helicopter

civvie -- civilian

claymore -- convex shaped above-ground anti-personnel mine

click -- kilometre

CO -- Commanding Officer

COIN -- counter-insurgency

Comrades Marathon -- 90 kilometre (55 mile) ultramarathon in SA, an annual event first held in 1921

connection -- friend, associate

CSM -- Company Sergeant-Major

cuca -- cuca shop, small local shops in Ovamboland
cutline -- bulldozed open swath through the bush, specifically the physical border between SWA and Angola

DB -- detention barracks ie. military jail, a brutal and harsh place
det -- detonator
dof -- lit. dim (Afr) not intelligent
dominee -- chaplain, reverend (Afr)
donga -- gully or ditch, usually formed by erosion
donner -- lit. thunder (Afr) mild swear word for hit, punch, slap, fall, same as 'moer' and 'bliksem'
doos -- lit. box, derogatory word for female genitals, used as a general insult (Afr)
Doug Domokos -- famous American motorcycle stuntman of the 1980s who held the world distance record for wheelying
draadkar -- lit. 'wire car' (Afr). An innovative toy made from coathangers and other wire by African boys everywhere

Echo tower -- One of a series of concrete water towers along the road between Ondangwa and the border post at Santa Clara

FAPLA -- military wing of the Angolan MPLA party, became the Angolan army in 1975
feel fuckall -- don't care
fire plan -- artillery barrage on predetermined targets or on a specific time schedule
Flossie -- SAAF transport aircraft, C-130 Hercules and C-160 Transall
FN MAG -- 7,62mm light machine gun carried by one man in each infantry section
fok -- Afrikaans version of 'fuck', used prolifically
fokkop -- fuckup

G

G3 -- Heckler und Koch 7,62 mm rifle made in Germany

gatvol -- lit. 'arse-full', fed up, annoyed (Afr)

gooi -- throw (Afr)

Grasshoppers -- casual leather shoes popular in SA in those days

grens -- border (Afr) in this context, a generic name for the operational area in SWA

grensvegter -- lit. 'border fighter', derogatory name for individuals who act like the hero or tell bullshit war stories

Grey Scouts -- horse-mounted infantry in the Rhodesian army

gunship -- armed helicopter, specifically Alouette III armed with a 20 mm cannon

gyppo -- cheat, weasel out of duty, do something less than complete or correct. Originally WW2 British slang for 'Egyptian'

H

hardegat -- lit. 'hard-arsed' ie. tough, no-nonsense (Afr)

HE -- high explosive

hoog voor -- lit. 'high front'. Ready position across the chest in which rifles were held when running (Afr)

houding -- lit. posture, looking and acting professional and sharp (Afr)

HQ -- Headquarters

hurl -- vomit

J

jaag -- race, ride at high speed (Afr)

jam stealer -- derogatory name for non-combatant, same as 'base wallah'

JL -- Junior Leader(ship), a one-year course at Infantry School in Oudtshoorn that produced platoon commanders and platoon sergeants

jol -- fun, enjoyable

'jou beurt is jou beurt' -- 'your turn is your turn' (Afr) fatalistic attitude

'julle sleg bliksems' -- 'you useless bastards' (Afr)

K

kaalgat -- bare-arsed (Afr)

kak -- shit (Afr), with 'fok' and 'poes' the most-used words in the SADF vocabulary

kapoen -- from 'kak en pampoen' (Afr), lit. 'shit and pumpkin'. Yellowish brown colour, specifically of berets the admin types wore

kas -- locker or cupboard (Afr)

kay -- kilometre

KDX 200 -- legendary Kawasaki 200cc dirtbike, light, nimble and powerful

klaar in/out -- signing in or out at a unit, drawing or returning equipment upon first arrival or final departure

klaarstaan -- stand to, usually at first and last light (Afr)

klap -- lit. slap (Afr) slang for something/one getting contacted in any way, hit, punch, clobber

'kokke kakke en klerke...' -- 'cooks, shits and clerks, fuck off and go do your thing'. Indicative of the general, but unspoken disdain front-line soldiers had for non-combatants

Koevoet -- Police counter-insurgency unit, extremely successful, feared and controversial

kos -- food (Afr)

kraal -- compound, a living area surrounded by a fence typically occupied by an extended family in rural Africa (Afr)

Kwêvoël -- 10-ton 6x6 truck chassis with a mine-proof drivers cab, an extremely capable vehicle

L

'laat hom lê waar hy geval het' -- 'let him lie where he fell'

lag af -- lit. 'laugh off', ignore or disregard someone (Afr)

Loot -- Lieutenant

M

mahangu -- pearl millet, a grain staple food in Ovamboland

mal -- crazy, wild (Afr)

meat pie -- MP, military police

middlemannetjie -- ridge between the two tracks of a tweespoor (Afr)

mieliepap -- corn maize porridge, a very popular African meal

min -- few, a little (Afr)

min days/min dae -- popular slang term for 'short time'

moer -- swear word with many meanings (Afr). Hit, punch, beat up, fall, crash, etc. Same as 'bliksem' and 'donner'

mopani -- a tree prolific in Ovamboland and Angola

mugu -- lowlife, dirtbag
MX -- motocross

N

nafi -- 'no ambition, fuckall interest' ie. unmotivated
NCO -- non-commissioned officer
Noddy car -- SA armoured car based on the French Panhard. Named after Noddy, an elfish English childrens' book character who drove around in a small car
NSM -- National Serviceman

O

oke -- from 'ou' (Afr) slang for male individual
omkeer -- 'about turn', drill command (Afr)
one liner -- the lowest NCO rank, Lance-Corporal
one-pip -- the lowest officer rank, Second Lieutenant
'Ons het hulle afgekoel....' -- 'We cooled them down, now we're warming them up'.
oom -- uncle (Afr)
opfok -- 'fuck up' (Afr) in this context, punishment by physical exertion like running up hills with sandbags or ammo crates. Could go on for many hours, depending on the transgression
Otavi -- small town in northern SWA where SWASPES was based
Oudtshoorn -- medium-sized town in the Karoo, the site of Infantry School where infantry leadership courses were conducted
Ovambo piele -- lit. 'Ovambo penises', tinned sausages

P

pap -- lit. porridge, in African context, a doughy staple food made from ground maize
paraat -- ready, motivated, militaristic attitude (Afr)
parabats -- paratroopers
PB -- plaaslike bevolking (Afr), local population
pelbev -- peleton bevelvoerder (Afr) platoon commander
pencil tiffy -- lit. 'pencil mechanic', ie. administrative personnel
PF -- Permanent Force
piel -- vulgar Afr word for male genitals. Used almost as prolifically as 'poes'

pip -- rank indicator for officers, lieutenants in this context

poes -- vulgar Afr word for female genitals. With 'fok', the most prolific words in SADF vocabulary. Used as a verb, noun, adjective or any other modifier

poes beret -- sloppy or silly looking beret, infantrymen were very particular about the shape of their berets

poes proppie -- lit. 'vagina plug', vulgar

pomp -- lit. pump (Afr) same as shag, bone, hump, boink, bang, fuck

pong -- stink

porky-pie -- lie, from Cockney slang

POW -- prisoner of war

Porra -- Portuguese

Potch -- Potchefstroom, a large town in western Transvaal where Equestrian Centre/Berede Sentrum was based

PT -- physical training, always guaranteed to be rigorous

pukka gen -- truth, real information. Old British slang that originated in India. Opposite 'duff gen', ie. bad or false information

punda -- supposedly a Zulu word for female genitals, used as slang for that, or just 'girl'. More plausibly, slang invented by the SA comedian Joe Parker

PW's poes plaas -- PW (Botha) the SA President's, plaas (farm), derogatory name for the SADF

R

R4 -- 5,56mm calibre infantry rifle, based on the Israeli Galil

Ratel -- very capable SADF six-wheeled armoured personnel carrier. Its many versions included the Ratel 20 with a 20 mm, and the Ratel 90 with a 90mm cannon

ratpack -- ration pack, a small box containing a 24-hour food supply

recce -- short for 'reconnaissance'. Also a general term for special forces, ie. Reconnaissance Commandos

rev -- from 'revolution', slang for mortar attack

roastie -- 'road rash', wounds resulting from crashing without protective gear

roman candle -- parachute not deploying properly

rondfok -- fuck around (Afr) in military context, to annoy, mess with, or chase troops around

Rooibaard -- Red Beard, an individual's nickname

rooi gevaar -- lit. 'red danger', the SA government's anti-communist propaganda campaign (Afr)

RSM -- Regimental Sergeant-Major, the highest ranking NCO in battalion-sized and bigger units

rustig -- calm, quiet, easy-going (Afr)

RV -- rendezvous, meeting at a predetermined place

S

SA -- South Africa

saak -- case (Afr) legal term, used to indicate disinterest or dismissal

SACC/SAKK -- South African Coloured Corps

SAI -- South African Infantry (Battalion), numbered 1-8

SAMIL -- South African MILitary trucks built in 20, 50 and 100 ie. 2, 5 and 10 ton versions

sappers -- combat engineers, finding and lifting landmines was their main function

Scope -- South African girly magazine from the 1970s and 80s

section -- smallest infantry unit, usually ten men

shona -- oshona (Ovambo) large open areas that filled with water in the rainy season and turned into shallow lakes

siel tiffie -- lit. 'soul mechanic', ie. chaplain

sitrep -- Situation Report, a daily status and position report to HQ

'skiet hulle' -- 'shoot them' (Afr)

Skiet Piet -- lit. 'Shooting Pete' (Afr) a derogatory name for Citizen Force soldiers, or 'campers' who tended to be trigger happy

skrik -- fright (Afr)

sleg -- bad, poor, useless (Afr)

slik -- steal

sluip -- sneak (Afr)

snotneus -- lit. snotty nose (Afr) common name for the M79 40 mm grenade launcher

soek-steek-stokke -- lit. 'search-stab-sticks'. Long rods used to probe for landmines in soft terrain (Afr)

spaar piel in die hoer huis -- lit. 'spare cock in the whorehouse' ie. unneeded or excess (Afr)

Spes -- slang for SWASPES

splab -- slang for saying something, usually when talking shit

spoor -- tracks (Afr), human in this context

spy, spies -- personnel of the Intelligence Corps. We didn't regard them as such after our dealings with the 2IC at 52 Bn

staaldak -- lit. 'steel roof' (Afr) the steel helmet all infantrymen wore during training, but never in the bush except on conventional 'externals'

steek -- lit. stab (Afr) in motorcycle context, riding at full throttle

storing -- jam or misfeed of a weapon, used as slang for 'problem' (Afr)

SWA -- South West Africa, became Namibia in March 1990

swap -- swapo, enemy combatant

SWAPO -- South West African Peoples' Organisation

SWASPES -- South West Africa Specialist Unit based at Otavi in northern SWA. Administrative unit for horse and motorcycle platoons, trackers, and dog handlers

SWATF -- 'SWA-tee-eff', South West African Territorial Force, the SADF- controlled and -equipped 'SWA army'. SWASPES was a SWATF unit

T

TB -- temporary base/tydelike basis, outward-facing defensive position taken up at night by units in the field

terr -- terrorist

tiffy -- slang for mechanic, usually anybody from the Technical Service Corps, they wore black berets

toppie -- father, old man (slang), same as 'ballie'

trap -- lit. step (Afr) slang for 'walk'

trek af plak toe -- lit. *'pull down, paste over'* (Afr) slang for stopping or ending something. At the shooting range, troops in the pit lower the targets, calculate the shooter's score and cover the bullet holes with small round stickers upon this command

trommel -- lockable metal storage trunk issued to every soldier

tweespoor -- twin track

two-pip -- Lieutenant

U

UNITA -- pro-Western faction in the Angolan civil war supported by SA and, initially, the USA and CIA

V

varkpan -- lit. 'pig pan' (Afr) stainless steel tray with indentations that served as troops' dining plates in messes everywhere
'verbroedering met offisiere' -- fraternising with officers
vleisbomme -- lit. 'meat bombs' (Afr), derogatory name for paratroopers
vrot -- rotten (Afr), slang for drunk

W

webbing -- harness worn by soldiers, with various attachments to carry ammunition and other equipment. SWASPES used nylon vest-type webbings, with magazine and grenade pouches in front and a general-use pouch in the back. The SADF's canvas webbing dated from the Second World War and was finally updated in the mid-1980s
white gold -- 'wit goud' (Afr) toilet paper, indicates how much it was valued
white phos/WP -- white phosphorous grenade, a hand grenade packed with phosphorous that, once burning, is almost impossible to extinguish. Extremely dangerous device, used to start fires or produce smoke
windgat -- lit. 'wind-arse' (Afr) cocky, overconfident. Usually ended badly
Withings -- lit. 'white stallion' (Afr), a highly capable mine-proof recovery vehicle based on a Samil 50 chassis

X

XR -- Honda XR 500 offroad motorcycle

Z

zol -- weed, marijuana, aka 'dagga'

101 Bn -- SWATF unit composed of Ovambo soldiers

2,4 -- 'two comma four' the 2,4 km weekly fitness test run, maximum time allowed was 12 minutes. Previously known as the 'one and a half' (miles). Infantry always ran with helmet, webbing and rifle. After a few months, over 10 minutes was not considered good enough

2IC -- Second-in-command

32 Bn -- Very experienced and feared SADF unit made up of former Angolan FNLA fighters, led by SA officers and NCOs

3 SAI -- 3 South African Infantry Battalion based at Potchefstroom

40 Days -- an important landmark for National Servicemen of all ranks: 40 days before finishing their 2-year stint in the SADF. It has all kinds of religious connotations, but in SADF context was based on the Cliff Richard song '40 Days' which was played loudly and repetitively during the parties that ensued

44 Para -- 44 Parachute Brigade

52/53 Bn -- for command and control purposes, the Operational Area in SWA was divided into Sectors 10, 20, 30. Sector 10 which controlled Ovamboland was sub-divided into 51, 52, 53 and 54 Battalions

61 Mech -- 61 Mechanised Infantry Battalion, a Ratel-mounted conventional warfare unit that mostly conducted 'externals'